John O'M. Bockris has been Distinguished Professor of Chemistry at the Texas A&M University, USA since 1983. Born in South Africa, he graduated from Imperial College, London and continued his research work there until emigrating to the USA in 1953. After nearly two decades as Professor of Chemistry at the University of Pennsylvania, he was appointed to the Chair of Physical Chemistry at Flinders University in South Australia, where he co-founded the International Association for Hydrogen Energy in 1975. Professor Bockris has received many international prizes for his work and is author of numerous books and articles.

T. Nejat Veziroğlu is Director of the Clean Energy Research Institute at the University of Miami, USA, and editor-in-chief of the *International Journal of Hydrogen Energy*. He was born in Turkey and graduated from the Imperial College, London. He joined the University of Miami Engineering Faculty in 1962 and held several posts before becoming Director of the Clean Energy Research Institute. Dr Veziroğlu organised the first international conference on hydrogen energy in 1974, and co-founded the International Association for Hydrogen Energy in 1975. He has written many scientific articles, lectured throughout the world and has received several international awards for his research.

Debbi Smith became an anti-pollution activist in 1968 when, as a school girl, she wrote to Governor Ronald Reagan to protest about air pollution in the Los Angeles Air Basin where she then lived. In 1981, she became an Assistant, and since 1985 a Consultant, at the Hydrogen Research Centre at the Texas A&M University. She is also Consultant for the Centre for Electrochemical Systems and Hydrogen Research, a Lobbyist for hydrogen research and development and Manager of the National Hydrogen Association in the USA.

GW00481871

SOLAR HYDROGEN ENERGY

The POWER to save the Earth

JOHN O'M. BOCKRIS
AND
T. NEJAT VEZIROĞLU
WITH DEBBI SMITH

ILLUSTRATED BY HEIDI WEISS

An OPTIMA book

First published in 1991 by
Macdonald Optima, a division of
Macdonald & Co. (Publishers) Ltd

A member of Maxwell Macmillan Pergamon Publishing Corporation plc

British Library Cataloguing in Publication Data
Veziroglu, Professor T. Nejat
    The solar-hydrogen energy solution.
    I. Title   II. Bockris, Professor John O'M
    III. Smith, Debbi L.
    333.79

    ISBN 0-356-20042-6

Macdonald & Co. (Publishers) Ltd
Orbit House
1 New Fetter Lane
London EC4A 1AR

Typeset by Leaper & Gard Ltd, Bristol
Printed and bound in Great Britain by
The Guernsey Press Co. Ltd, Guernsey, Channel Islands.

CONTENTS

PREFACE

Gradually, inexorably, the levels of pollutants and carbon dioxide in the atmosphere are increasing.

The resulting global warming, acid rains and pollution are seriously damaging the biosphere of the earth – the only planet known to be hospitable to life. The consequences are immense, and largely, very undesirable indeed for the people of this planet, as well as for its flora and fauna.

The increases in carbon dioxide and other pollutants such as carbon monoxide, oxides of sulphur and nitrogen, hydrocarbons and soot, are due to the fact that we obtain our energy by burning oil, natural gas and coal. We have to stop this dependence on fossil fuels, which – each day it continues – puts even more carbon dioxide and pollutants into our atmosphere and makes the situation worse.

In the recent past, writers could only describe the effects of global warming, acid rains and pollution, and then wring their hands in despair. Now we are offering a solution to the world's dependence on fossil fuels, and a way to reverse their damaging effects – through the solar-hydrogen energy system.

This book is the first to give a clear and lucid description of WHAT CAN BE DONE to solve the problem – and indeed how you, the reader can be part of the solution.

John O'M. Bockris
T. Nejat Veziroğlu
Debbi L. Smith

ACKNOWLEDGEMENTS

In producing this book, we had the support and assistance of many people. We gratefully acknowledge the interest and suggestions of our immediate colleagues and research workers, including Nigel Packham and Jeffrey Wass of Texas A & M University; Robert J. Adt, Jr, Frano Barbir, Jerome Catz, Akira Mitsui, Harold J. Plass, Jr, Sherif A. Sherif and Michael R. Swain of the University of Miami; and Peter Hoffmann and Curtis A. Moore of Washington, DC. We would also like to acknowledge our very talented artist, Heidi Weiss, for her creative and humorous illustrations.

We also wish to extend our sincere appreciation for typing and retyping the manuscripts to Donna G. Pressley and Jainnie Leighmann.

And, last but not least, our thanks go to our spouses for their support and encouragement throughout the production of this book.

John O'M. Bockris
T. Nejat Veziroğlu
Debbi L. Smith

PART ONE

I
INTRODUCTION

More and more, in the media and in general conversation, we are being made aware of the fact that our atmosphere is being polluted and spoiled by actions which are beyond our control, and yet in which we play a vital part. For example, we have all heard of the greenhouse effect, the fact that the world is gradually getting warmer and that the seas will rise increasingly with this warming as the polar ice caps melt. But there are other things which are going on. The rain, usually the balm to follow a great heat, is becoming increasingly acid – fish are dying in the lakes; forests are dying all around us; some buildings are crumbling under the assault of the acidified rain. And human health is suffering from the increase in environmental pollution.

With all this, governments treat the assault on the atmosphere with promises and band-aid measures. The main promise is to 'study the effects', whereas – as is contended in this book – the causes of these problems are well known, and have been for some years.

Occasionally, as with recent announcements from various governments, remedial measures are promised: the amount of sulphur might be reducible in our fuels; factory chimneys will be made taller to try to take the smoke away from towns. But

TALLER FACTORY CHIMNEYS TO TAKE SMOKE AWAY FROM THE TOWN

the basic problem is not mentioned – we need to take the carbon out of the carbon-containing fuels that we now use, so that they will not produce carbon dioxide, for it is this gas that causes global warming.

If we think of these fuels as chemical compounds, they are made up of molecules containing two elements, carbon and hydrogen. But the truth is that, of these two elements, only the hydrogen is necessary for the release of energy as it is the hydrogen in the fuel which combines with the oxygen in the air upon combustion (and it is combustion that produces the heat energy which we use not only for warmth but for generating that most versatile of energy carriers, electricity) to form harmless steam and water. However, since carbon (as well as other chemicals) is also present in the fossil fuels we use today, it is also found in exhaust fumes or chimney emissions after combustion and is released as the gas carbon dioxide or CO_2.

Over many years, huge quantities of CO_2 have been released into the atmosphere. This CO_2 acts like the glass in a greenhouse, trapping the sun's heat and causing the earth's atmosphere to warm. It's not too great an effect right now, maybe $\frac{1}{2}$°C or so, but as time goes on the temperature will increase. Recent summers have indicated that the time of significant increase may have arrived, and that as long as we continue to use our present carbon-containing fuels, we can expect nothing from the future but choking rising heat. So, in this book we will attempt to explain the origin of this heat, and why and how it is produced and what will happen to us if we let the carbon-containing fuels go on being used to power our transport systems and factories. This book will also take us far away from this dismal scenario to a brighter one, showing how it is not necessary to use such fuels; other, non-carbon-containing fuels can be made available. It is all a matter of action being taken by governments, but this will only happen if we, the people, insist on change. What opponents to change will tell you is that the change will be too expensive, but the truth is it doesn't have to be. We have to phase in the new clean fuels gradually, so that we do not send the world's economy into shock.

The essential message of this book, then, is simple. If we go on using carbon-containing fuels we will suffocate ourselves with heat. Furthermore, it isn't necessary to do so. Carbon-free fuels are available, ready for massive manufacture, but the people have to demand that they be produced. The real power for change lies in your hands.

2
HOW WE GET OUR ENERGY TODAY

There was never a grand plan for obtaining energy for the human race to utilise – it just happened. As our use of energy progressed, so our development progressed – from the gathering of dung of undeveloped societies to the energy technologies of oil, natural gas and coal we have today. Energy provides us with basic comforts such as light and warmth, but has also enabled us to travel over vast distances by air, land and sea; to build and operate factories; to manufacture labour saving devices from washing machines and computers to electric toothbrushes. But while our energy usage was progressing and becoming ever more complex, no one could foresee the harm this would inflict on our planet.

In this chapter, we give a very brief overview of what energy is and how we get our energy today. But first we will try to define energy. This isn't all that easy; it's like trying to describe a will o'the wisp. We can't really point to it or weigh it, but we know energy is there.

- For instance, heat is energy in a sense, not only in the obvious way as being a source of warmth. Heat can also be used to boil water, producing steam which drives turbines to generate electricity. So we can see that heat is energy.
- It can be difficult to bend metals, but when two cars collide, all sorts of bits of metal get bent. The collision is the energy of motion, or kinetic energy, and the force behind it is enough to break up a car.
- Another kind of energy everybody knows about is stored energy. Coal stores energy, because when we burn it with

the oxygen in the air it gives us the heat that boils the water that drives the turbine that makes electricity.

So now you know the three basic forms of energy – heat, kinetic and stored. But there are two more types of energy that relate to the other three – chemical energy and electrical energy.

- Chemical energy is the kind of energy stored in coal, and is only released when a chemical reaction takes place – For example combining oxygen and coal in the presence of heat, or what most of us simply know as burning.
- Electrical energy is another kind of energy. When some chemical reactions occur, electrons are released and create an electric current. A battery is one source of electricity produced in this way. Electricity can also be generated by spinning a coil of wire in a magnetic field. This is how a dynamo works. Conversely, electrical energy can make metals magnetic, so that it is useful for magnetic motors – the kind found in lifts, for instance.

There are thus many forms of energy, and as human societies have developed, so have the ways we obtain and use energy. The main problem facing us now, as you will see in the next few chapters, is that most of our fuels, which provide the energy we need, contain carbon.

FROM HORSEPOWER TO NUCLEAR POWER

Prehistoric people had only the energy of their muscles to accomplish work. The subsequent domestication of animals, such as horses to pull wheeled vehicles, was the first important means of supplementing the energy of our own muscles. So we began to measure our rate of energy by the number of horses it would take to perform the task – horsepower.

The Newcomen engine
Thomas Newcomen was bent on making the horse and cart a

THE NEWCOMEN ENGINE

thing of the past; he was the inventor of the first practical steam engine, just before 1712. His engine was a one-piston affair, driven by heating a boiler with coal to produce the steam to drive the engine, and was used to pump water out of mines. Not only was his invention useful for the miners, but it was the forerunner of all subsequent steam engines.

But what Mr Newcomen had no way of knowing was that his brilliant invention, and all the ones to come after, would emit carbon dioxide into the air. Over the years, as the coal and wood and later oil and gas were burned, carbon dioxide and other chemicals would gradually accumulate in the atmosphere and trap the heat that now threatens the world's population with what is known as the greenhouse effect (see Chapter 4).

From pumps to trains

It's obvious that when coal or wood is burned we get heat, but how does this heat – one form of energy – get converted to useful energy, mechanical energy, to do things?

In Newcomen's engine the idea was that water was heated until it was boiling and then the steam was let, under pressure, into a cylinder containing a   piston. When steam enters a cylinder at such great pressure it immediately tries to escape again, and the only way it can do this is by pushing the piston and moving the shaft of the pump. And if the piston was connected to a wheel instead of a pump, the wheel would turn. That's how the first steam locomotive *The Rocket*, invented in 1814 by George Stephenson, worked.

From pumps to trains

Out of steam and into internal combustion

In Newcomen's steam engine the fire was outside the boiler, the flames licking the casing of the boiler itself, heating the water inside to produce the steam. This was wasteful as much of the heat from the fire escaped into the air, and a bit

Out of steam and into internal combustion

cumbersome, but it didn't take long before a more compact engine was devised – the internal combustion engine, the basis of the engine used in the cars we drive today.

However this engine is a little more complicated than a steam engine because the heat energy has to be produced quickly in a series of pulses or bursts – one pulse for every time the piston goes through a stroke. The only way to do this is not to bring steam to the piston, but to bring together two gases which, when combined, form a hot exploding mixture that expands and pushes the piston away, thus driving the wheels. It is the repetition of this explosion, about 50 times a second, that drives your car.

In an internal combustion engine the two gases that combine are petrol vapour and the oxygen in the air. A spark causes them to explode and the exploding gases, being much hotter than the cold vapour of petrol and oxygen, therefore try to occupy a greater space (because gases expand on being heated). And again, the only way they can do this is to push the piston out of the way; the piston then moves the crankshaft of the car, turning it, and the rotating crankshaft drives the wheels.

Electricity

Have you ever thought about how many devices in your house use electricity? Light bulbs, the television, the stereo and refrigerator are obvious examples. But what about all the others – the coffeemaker, washing machine, vacuum cleaner, hairdryer, blender, beater or food processor. At the office or workplace electrically operated computers perform a number of tasks, and each of the computers usually has an electrically operated printer. There may be electric typewriters, postal machines and electronic scales. Mundane factory work, once done by humans, is now often performed by electronic robots. If your paycheque has a facsimile signature, chances are it was put through an electric machine called a protectographer to make that signature. The majority of these items mentioned above have one thing in common; they are either run by or are dependent on an electric motor.

A man named Nikola Tesla invented the alternating current motor – the precursor to the electric motor we have today – and thanks to Mr Tesla's inventive mind in 1888 it is unlikely that our present need for electricity will decrease. This is why we have to be concerned about how our electricity is produced.

Today, as in the demonstration projects of the 19th century, electricity is mostly produced from coal. In 1866 Johann von Siemens invented the dynamo that was able to convert the heat energy, derived from burning coal, into electricity; the dynamo used the coal-produced heat to heat air, the hot air expanded with enough force to push a piston, the movement of the piston turned a rotor, and the rotor set a coil of wire spinning fast enough in a magnetic field such that it produced an electrical current. The principle of the dynamo is still one of the chief methods of production for the electricity that we use to drive all the electric motors on which we are so dependent, not to mention the gadgets that don't have motors, like light bulbs and television sets.

Electricity is a remarkable and versatile transporter of energy, but it shouldn't be necessary to harm the environment for the sake of our washing machine or the equipment at the office. We therefore have to find a method to produce electricity cleanly.

Nuclear energy — heat and electricity

Nuclear energy, once thought to give us an inexhaustible supply of electricity, has had a change of image since its conception in the late 1940s and early 1950s. Now, instead of being incredibly cheap (as was originally thought), it has turned out to be very costly because of the expense involved in constructing, operating and maintaining the nuclear power plants.

To understand how nuclear energy can be harnessed we must first learn a little bit about the structure of an atom. An atom is much like our solar system, but on a minute scale. There are two parts to the atom. The nucleus is the central part of the atom – the 'sun' – around which the electrons rotate, much like the planets do around the sun. It's these electrons and how they arrange themselves that produce chemical energy. The difference between chemical and nuclear energy is in the amount of energy we can get for the same weight of fuel, for when the energy within the nucleus of the atom is released it is about a million times greater than the energy you can get from the electrons. Now you can see why people once thought electricity from nuclear energy would be so cheap.

After people discovered they could tap this energy within the nucleus of the atom and use that energy for something other than weapons, nuclear reactors were built. Nuclear reactors use the same principle in producing electricity that steam and internal combustion engines do; the nuclear energy is used to heat a liquid (usually water) to produce steam that drives the turbines that turns the rotors of the generators that produce electricity. The important feature of nuclear energy is that the fuel – uranium – contains so much more 'stored' energy than either coal or oil, so that comparatively tiny quantities of it are needed. But the major problem with nuclear power is, of course, radiation which is known to cause cancer and is therefore extremely dangerous.

THE GOOD NEWS AND THE BAD NEWS

We have shown the basic ways in which we obtain energy now. There are two good things to say about this: the energy is cheap, and we have all the systems and infrastructure in place to continue using oil, natural gas and coal.

The bad news is that although our energy is cheap it is not inexhaustible, and, as you will see in the next chapters, using fossil fuels is having a devastating effect on our environment and our health.

But let's examine the first of these two points in more detail. The fossil fuels we use were formed by rotting trees, plants and animals. The trees and plants (which also fed the

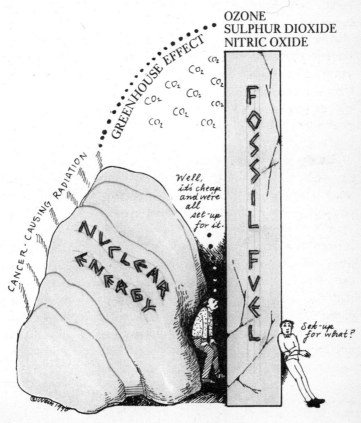

Between a rock and a hard place

animals) were in turn dependent on the sun's energy for the process of photosynthesis. The fact that this energy is already packaged and stored, trapped by nature over millions of years naturally means that it's cheap; you might say that we had a grant – a kind of stored capital – to begin and develop our civilisation. However the time has now come to stop spending our capital with no regard for tomorrow; we can't wait until it's all gone, and then cry out that we need more. We have only a relatively few more years of stored fossil fuels, and if we continue to burn them it will cost us the environment and our health – not a wise investment.

What we must learn to do instead is to invest our profits from the use of this stored energy in the future. We need to invest in a new clean energy system. Once our investment is complete, our future secure, then we need to save our stored capital – our fossil fuels – and use the small amounts required to make things like synthetic fabrics, ink, aspirin, and all other items that can't get manufactured once the fossil fuels do run out.

3
THE STORY OF DOOM – POLLUTION

In this chapter we give a very brief overview of several of the unnecessary things that are happening to our biosphere.

The most important is undoubtedly the blanket of CO_2 that is causing the greenhouse effect – the warming of the planet due to trapping solar heat in the atmosphere. But where does the carbon dioxide come from?

OUR FUELS CONTAIN CARBON

When people stopped using the horse and cart in exchange for the early belching and backfiring cars, they first had to rely on the refining of oil to make fuel to power the automobile. There weren't as many petrol stations then as there are now, and tanker trucks weren't designed like the ones we have today. In other words, driving and refuelling weren't as convenient then as they are now.

Today, we have the convenience of being able to live far away from our jobs if we choose to, and travel to them every day; but an invisible price is being paid for the amount of petrol (and natural gas) we use in ever increasing amounts. That same price is paid in the use of coal to produce electricity, and the reserves of coal that will be used as the source of synthetic fuels when the oil runs out. For these fuels – the principal fuels we use at the present time – all contain carbon, and hence when they are burnt or ignited with oxygen in the air form carbon dioxide, CO_2, the greenhouse gas.

For many years, CO_2 was regarded as a benign gas as long as it was not inhaled in great amounts, for it is not poisonous and does not cause cancer. Furthermore, CO_2 is released into the atmosphere in the natural world (sometimes in vast quantities, such as during volcanic eruptions), but various processes, such as photosynthesis in which green plants use carbon dioxide to produce sugars, kept levels of CO_2 in balance. Indeed, plants, and therefore life, could not survive without CO_2. So, what did it matter that a little CO_2 was added to the atmosphere every time a car's wheels turned?

But gradually, as the decades went by, but particularly since the 1970s, scientists began to realise that the massive increase of CO_2 in the atmosphere emitted by our cars and factories does cause harm to the world we live in – indeed, very grave and increasing harm. And by the 1980s, many more people had realised the dangers of the increasing levels of CO_2 in the atmosphere, caused by our use of carbon-containing fuels. Unfortunately, by the time we understood the far reaching environmental effects of too much CO_2, our fuel and power systems were well in place and no one wants to change them.

NITROGEN AND OXYGEN COMBINED

But there is another malady upon us.

Our principal method of obtaining the energy for our technologically complex society is to produce heat through a combustion process. We obtain this heat by taking carbon-containing fuels and burning them with air – a mixture of oxygen and nitrogen – the by-products of this process being carbon dioxide (CO_2) and steam (H_2O). But when we do this another chemical reaction takes place, which we do not often speak about. The nitrogen in the air combines with the oxygen to form a gas known colloquially as NOX. This NOX in fact represents two separate gases, nitric oxide and nitrous oxide, indicated by the chemical formula NO and NO_2. Both of these gases exist in an uneasy mixture together, so we do not bother in most of our discussions to distinguish between them. We simply call them NOX.

Unlike CO_2, NOX attacks us directly as it turns into acid in our lungs when we inhale polluted air. The main physiological damage this causes is to eat away at the lung's spongy structure so that it gradually loses its capacity to absorb oxygen from the air, so less oxygen is passed into the bloodstream. This may eventually lead to death, but before that happens the person will probably suffer from chronic shortness of breath and lung and bronchial illnesses such as

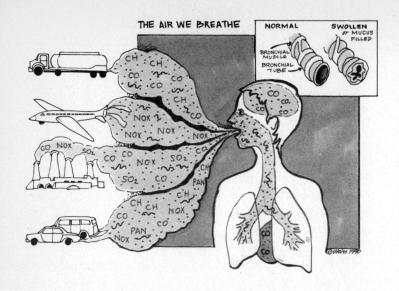

emphysema. In the South Coast Air Basin near Los Angeles, the main health hazard is now NOX, or PAN the acid-containing component in petrol (see page 20), rather than CO_2.

So burning our carbon-containing fuels not only causes the atmosphere to warm up and the planet to become overheated, but it also produces noxious substances which are harmful to our health: bronchial tubes fill up and lung diseases develop. Furthermore, as the amount of NOX in the atmosphere has increased, so we have slowly been killing our trees; the trees gradually lose their ability to fight disease and they slowly die.

THE HARMFUL EFFECTS FROM THE SULPHUR IN COAL AND OIL

Not everyone knows that coal and oil, which are the primary types of carbon-containing fuels, also contain some sulphur. Oil has less than coal, but it's still there, so if we burn oil or coal (and we burn a lot of coal to produce electricity) then sulphur is produced at the same time.

This sulphur is emitted in the form of sulphur dioxide. Sulphur dioxide causes similar problems to human health, especially to the respiratory organs, as NOX. And by the time

it gets into the atmosphere it oxidises further to another compound called sulphur trioxide (SO_3), and finally combines with rain water to form sulphuric acid. Therefore, when the rain comes down, the rain itself is acidic.

Dangers of relying on coal

But you know what everyone says about coal – it's our great fuel reserve. There are supposed to be billions of tons buried in the earth, and if the going gets tough in the Middle East and we can't get fuel from there, we can still dig down and get the coal out of the ground. But it's a mistake to believe we can live on it; depending on coal would be far worse from a polluting point of view than depending on oil, which is bad enough. Exchanging coal for petrol would increase the pollution difficulties enormously; coal adds to the carbon dioxide problem and, as explained above, gives off sulphur to form sulphuric acid. There is also the danger of fly ash, increasing the number of particles in the atmosphere and causing smog.

Even if we did have large amounts of coal, more than 100 years' worth – which is what many experts say we have (some say as much as 500 years) – it wouldn't be worth burning it because of the tremendous pollution it would cause.

OTHER POLLUTANTS

Another pollutant, which comes from burning petrol, is called peroxyacylnitrate, nicknamed PAN. PAN is a very nasty substance. It causes swelling and irritation of the larynx (the voicebox in the throat) and irritates the eyes. If inhaled sufficiently, you experience a burning, which may make it difficult to breathe.

THE LINE-UP

The next pollutant we want to name is benzopyrene, and it may be worst as far as long-term damage to the body is concerned. It has this bad reputation because it was shown to be the chemical in cigarettes that causes lung cancer. To what extent it may cause cancer when inhaled from diesel fumes isn't known yet, but we do know it's there.

Another problem with burning petrol, diesel or any other fuel that contains carbon is that carbon monoxide is produced. Now carbon monoxide is very different from carbon dioxide. Carbon monoxide is poisonous; carbon dioxide is not. Carbon monoxide is what asphyxiates people who inhale vehicle exhausts without enough ventilation. Carbon monoxide is found in smoggy areas too. When we inhale it, it causes headaches, dizziness and confusion. The carbon monoxide tricks red blood cells into thinking they have oxygen from the lungs, so instead of taking the oxygen to

WHAT POLLUTION CAN DO...

our brains and heart, the red blood cells feed carbon monoxide to these vital organs instead.

The last pollutant is one which we will discuss in more detail in Chapter 7 – smog. Smog is a dirty kind of fog, and is the cause of a great deal of trouble. For example in the South Coast Air Basin of Southern California, where it causes haze and glare, straining and tiring the eyes. Sometimes it's thick enough there to hinder visibility, according to Edwards Air Force Base, where the Space Shuttle lands.

Because smog contains all of the above mentioned pollutants, and because no large city is immune from smog any more, a significant percentage of our population suffers physically from unhealthy air. Do you?

OTHER EFFECTS OF POLLUTION ON THE ENVIRONMENT

The greenhouse effect, smog, effects on human health – all these problems have already been mentioned, but there are other effects of pollution. Some of these may surprise you.

Pollutants in the air can kill trees, damaging whole forests. This hapens because the pollutants clog the tiny pores in the leaves, stopping the trees from photosynthesising (the process by which plants manufacture their food) or respiring (literally breathing – this is the process by which plants break down food to give them energy). Another product of our polluted atmosphere is acid rain (see Chapter 5). We have already explained how sulphurous gases produced by burning coal and oil combine with rain water to produce dilute acid. Unfortunately, this affects trees and other plants, burning their leaves and dissolving minerals in the soil. The plants are unable to filter out these minerals, some of which are harmful to the plants. The most worrisome prospect is that the soil itself may become too acid to grow the vast amounts and variety of crops we depend on. These are all damaging effects we can see and measure now. What does the future hold if we do nothing to curb acid rain?

WHERE DOES THE POLLUTION COME FROM?

In the past it has been standard practice for governments to say that no one knew for sure where the pollution was coming from. Numbers of studies were done, and this research was a reason for postponing the spending of more money on alternative energy and cleaner methods of producing electricity. But now we certainly have more details about where pollution comes from than we did in previous years, and the main point stands clear: our air pollution comes mostly from the burning of our present fuels – coal, natural gas, oil and the refined products from oil. These fuels inject smoke, smut and a number of other damaging compounds into the atmosphere. Power plants that produce electricity, if they're not nuclear, use coal, oil or natural gas to generate the power. What gets discharged into the atmosphere often contains a great deal of sulphur, which will eventually cause acid rain.

Another large source of pollution comes from burning wastes in incinerators. If the burning occurs without any attempt to collect the dust and debris, these rise high in the air and stay there, becoming part of the dirt contained in smog, eventually falling as part of the dirt and grit more often found in cities and urban areas than in the country. With incinerators we also have to watch out for chemicals released from our plastics, packaging and other disposable articles; these chemicals stay in the air along with the chemicals released from our factories and vehicle exhaust pipes.

It is plain that all these incinerators, manufacturing chimneys belching smoke and debris, the exhaust pipes of cars and the smokestacks of electric power plants are sources of pollution. In fact, they were never planned not to pollute; in the days when they were first built, the people thought the pollution would be so minimal that it could easily be dispersed in the atmosphere with no damaging effects.

WHAT IS THE COST OF POLLUTION?

Dirt, damage, overheating, illness, smog, the eventual decline of agriculture, rising seas, the flooding of cities – all these are the horrors to come. But right now (January 1991) the cost of the pollution which carbon-containing fuels cause is around an extra US$1.30 per gallon of petrol.

These extra costs are silent; they do not come forth and shout for themselves, but they are nevertheless there. We are paying for them all the time, though we do not pay for them at the petrol station. The cost of the fuel we buy yields splendid profits to the oil producers, but it does not include the cost of the damage that the use of the fuel is doing to us. There are some of us who think that the real cost of fuel should be taken into account; people will then be encouraged economically to use carbon-free fuels, so that eventually the health-damaging carbon-containing fuels will no longer have a market.

4
THE GREENHOUSE EFFECT

What exactly is the greenhouse effect that everyone is talking about?

A greenhouse is designed with a transparent roof and side panels, normally made of glass, that allow sunlight to reach the plants kept inside, keeping them warm so that, in the middle of a frigid winter outside, a temperate or tropical climate keeps the plants lush and green inside. And the greenhouse effect for the earth works the same way. Certain gases in the atmosphere act as the transparent roof, the most abundant of these 'greenhouse gases' being carbon dioxide; the others are methane, carbon monoxide, hydrocarbons and chlorofluorocarbons (CFCs). These gases are transparent, so the high-temperature radiation from the sun (sunlight) passes through the atmosphere and reaches the earth, but the low-temperature radiation from the earth (heat) is prevented from escaping by the greenhouse gases. This is because the energy is absorbed by the carbon dioxide, and as a result this heats up the atmosphere, which in turn causes the whole planet's warming.

Our planet has to have some carbon dioxide in its outer atmosphere to make it warm enough for life. If we look at our closest planetary neighbours in the solar system – Venus, whose atmosphere contains more carbon dioxide than earth and is consequently very hot, and Mars, whose atmosphere doesn't contain as much carbon dioxide as ours and is consequently very cold – we see that the earth contains just the right amount for life as we know it. Therefore we don't want to get rid of all the carbon dioxide, lest we become like

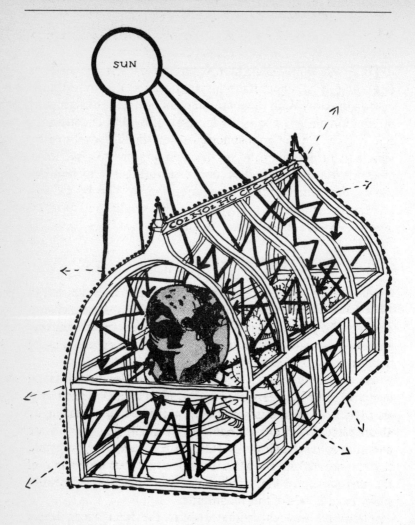

Mars, but the more we pollute our air the more we become like Venus, and we don't want that either.

The world's temperature has increased ½°C (about 1 °F) since the Industrial Revolution. Why? For thousands of years, wood has been the main source of fuel for mankind – for many people it still is. Later, in some societies where it was available, coal replaced wood as the primary fuel because more heat is produced by coal than by the same amount of wood. Then, early in this century, oil began to replace coal because it was more convenient to transport, store and use.

With resources of oil fast depleting, natural gas is now becoming a more important energy source.

However, wood, coal, petroleum and natural gas all have one thing in common: the main by-product of combustion is carbon dioxide. With the increasing energy needs of a growing population and economy, the increase of fossil fuel consumption has caused the atmospheric carbon dioxide level to rise, which in turn causes the earth's temperature to rise. If we continue to use fossil fuels to meet the world's energy requirements, the planet's mean temperature could rise by 5°C (9–10°F) by the end of the next century. This would spell disaster, as we will see in the following sections.

RISING OCEANS AND KILLER HURRICANES

The increase in the earth's temperature has already started to melt the ice caps at the North and South Poles. In addition to melting the ice caps, the higher temperatures are causing glaciers and snow lines to retreat. These in turn are causing the ocean levels to rise.

Measurements from 243 locations around the world, taken over a period of 30 years, show that the sea level has risen by 15 cm (6 inches). It is estimated that now the oceans are rising by about 1 cm (a little less than half an inch) per year. If we continue to use fossil fuels at the present rate, it is projected that the oceans will rise 1.8–2.4 metres (6–8 feet) by the end of the next century. Think of all the cities at the coast that would suffer from floods of this magnitude.

There are various estimates as to the final ocean levels when the 'melt down' is complete, but according to more conservative estimates they will be 6–7 metres (20–24 feet) higher than today. These increases in ocean levels would flood all the coastal cities, where a large number of the earth's population lives. The coastal plains, which are the most fertile lands in the world will be flooded. When we consider the fast growing global population, doubling every 35 years, it becomes clear that the ocean rise will result in an extreme shortage of habitable and agricultural land.

For example, it is estimated that more than half of Bangla-

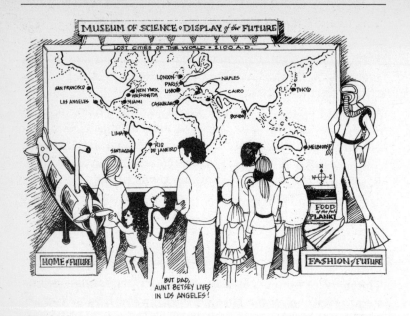

MUSEUM OF SCIENCE ◦ DISPLAY of the FUTURE

LOST CITIES OF THE WORLD ◦ 2100 A.D.

BUT DAD,
AUNT BETSEY LIVES
IN LOS ANGELES!

desh's habitable land will become unusable. The Nile Delta, the most densely populated area of Egypt, including Cairo and Alexandria, will be under water, leaving more than 30 million people homeless. Three-quarters of the state of Florida will join the intercoastal waterway. The Maldive Islands in the Indian Ocean will disappear off the map, and Venice's problem is not that it is sinking, but that the sea is rising.

The most catastrophic version of this argument brings in a concept called positive feedback. A great deal of CO_2 is dissolved in the sea. If the temperature rises due to levels of CO_2 in the atmosphere rising, then the top 70 metres of seas will also heat up. This will release some of the dissolved CO_2, which will further add to the greenhouse effect. This will increase the temperature of the sea even more, releasing even more CO_2, and so on. This would be called the catastrophic greenhouse effect, and we must be very careful that it does not happen. If it does, then this once benign pollutant could become the greatest disaster that has happened to us so far.

There have been few estimated costs of the resulting loss of cities and land. They would probably run into millions of billions of dollars. There have been some suggestions made that a high dam could be built around all the continents to protect them from the rising seas, like the dykes of the Netherlands. But this dam would have to be 9–10 metres (30–35 feet) high, and of course large pumping stations would be needed to pump all the world's rivers over the dam and into the oceans. Conservative estimates indicate that US$70 trillion (1990) would be required to build the dam; then add the cost of continuous pumping of all the rivers up over the dam, and you can see that the cost of the energy involved would be prohibitively expensive. It is unlikely any insurance company would insure such a dam; a disaster like a major earthquake or hurricane could crack the dam or cause a loss of power at the pumping stations, flooding immense tracts of land.

But that's not the end of it. An increase in the sea's surface temperature of 1°C (2°F) will decrease the minimum sustainable pressure in hurricanes by 15 to 20 millibars. Consequently, an increase of a few degrees could cause a substantial increase both in the number and severity of hurricanes. For example, Hurricane Gilbert of 1988, with a pressure of 885 millibars (1 millibar = 0.001 atm) and winds up to 330 km per hour (200 mph), was the most severe storm in recorded history; millions of people were left homeless, several hundreds were killed, and property damage was estimated at more than US$10 billion. Although any particular hurricane cannot be attributed to global warming, the frequency and ferocity of hurricanes (and, for that matter, typhoons and tornadoes) could be expected to increase with a greenhouse effect-induced warming.

DROUGHTS AND FLOODS

The greenhouse effect is also causing other climatic changes. With the rise in the earth's temperature, the rate of evaporation from the oceans, rivers, lakes and plants will also increase. Because of the balance of nature, more evaporation leads to more clouds and rainfall. However, as observed recently, wind patterns are changing, resulting in cloud precipitation pattern changes. This is causing drought in places where historically there was normal precipitation.

Droughts in turn can impoverish agricultural production – crops and livestock – and endanger wild flora and fauna. They can also reduce the quality and quantity of irrigation and potable water. Many experts believe that the drought in the United States in 1988 was a direct result of the greenhouse effect. It has been estimated that during this drought US$5 billion worth of cash crops were lost – all eventually paid for by the consumer in higher prices at the market.

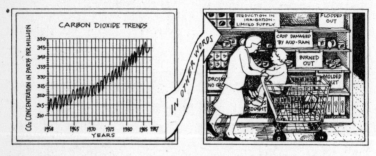

Source: Keeling, 1983; Komhyr et al., 1985; Conway et al., 1988, Rotty, 1987.

Unfortunately, when droughts occur in one part of the world, floods happen elsewhere to balance the water cycle – and that is precisely what happened in 1988. While droughts were occurring in parts of the United States and in central and western Africa, there were devastating floods elsewhere, such as those in Bangladesh. Three-quarters of the habitable part of Bangladesh was covered by flood waters that took several weeks to recede. Most of the country's cash crops were ruined. Many homes were destroyed or rendered uninhabitable. Innumerable bridges and large sections of roads were washed away. It will take an already impoverished Bangladesh years and large financial investments to recover from this disaster, even if another one does not hit them in the meantime.

The droughts and floods of 1988 are examples of the result of a small increase in the earth's mean temperature. If we take this into the future, when the increase in temperature will not be measured in fractions of degrees but several degrees, the enormity of the problem becomes apparent.

THE HIGHER THE HEAT, THE LOWER THE POWER

All electric power generation requires water; water is used for cooling and condensing the used steam in thermal power plants (those that run on coal, oil, natural gas or nuclear), and it is also used directly as a source of energy for hydro-electric power generation. For example, in the United States, 210 billion gallons of water are used per day for cooling thermal power plants, and 3,300 billion gallons of water per day pass through hydro-electric power turbines to generate electricity.

During a year of average conditions, sufficient water for power production is available; however, during the drought periods normal power production can be affected significantly. Furthermore, when it gets hotter there can be an increased demand for electrical power (for example, to power air conditioning equipment), but because of water shortages caused by the heat, the capacity to produce it is much reduced.

The economic impact of drought on power production is significant. During the 1977 California drought, Pacific Gas and Electric Utility incurred additional costs of approximately US$400 million above their annual operating expenses of US$1.3 billion. These additional expenses resulted from a shortage of hydro-electric power that had to be replaced with more expensive thermally generated, imported power.

In summation, droughts effectively reduce both hydro-electric power and thermal power generation, while causing increased power demand. This in turn causes hardships to society and enormous economic losses.

HAMPERED RIVER NAVIGATION

Another effect of high temperatures and droughts is that the water level in rivers is lowered, and this can greatly hamper river traffic and increase transportation costs.

For example, Mississippi River barges carry about 15 per cent of the bulk freight shipments in the United States. During the 1988 drought the river reached its lowest level since record-keeping began in 1871. River traffic backed up as barges had to wait in line. Eventually, at the height of the drought, when the river fell below its acceptable limit, the barge traffic was stopped completely. The stoppage resulted in the loss of revenues of some US$750 million. Some of the bulk goods had to be transported by more expensive means, such as rail or truck, increasing economic losses further. A conservative estimate puts the total traffic-related loss caused by the drought of the Mississippi River Basin at US$1.5 billion. And of course, as temperatures rise, as the droughts become more widespread, the losses will become greater and greater.

THE HUMAN COST

We should also mention the loss of human life. When people go about their business as usual during unusually hot days and do not take precautions, cardio-vascular problems can occur, resulting in deaths of quite healthy people. As the number of record-breaking hot days increases as a result of the greenhouse effect, so the number of heat-related deaths will also increase.

During the hot summer of 1988 in the US more than 80 cities and towns for the first time in history experienced temperatures of over 38°C (100°F). The number of deaths reported by the press during the height of the heatwave was 175. The press also reported some 100 deaths in the Mediterranean countries of Spain, Italy and Greece during August 1988. It is expected that the heat-related deaths are much greater in Third World countries, but these deaths usually go unreported either because of lack of communication or because the difference between natural deaths and a heat-related death is not clearly delineated.

During the heatwave of late July 1980 in New York, the mortality rate on the hottest day was over 50 per cent above normal (elderly people usually suffer the most on these days). Studies show that mortality increases after the temperatures rise above a certain limit (for New York it is 33°C, 92°F). In the world today, roughly 5½ per cent of the total days exceed the threshold temperature during an average summer, but for a 4°C (7°F) global temperature increase (not yet reaching the 5°C, 9–10°F rise predicted by the end of the next century), over

Dear Kids,

33 per cent of the days in summer would exceed the threshold. For the world as a whole, this increase would mean 680,000 deaths during the summer attributable to high temperatures.

Is it worth this to keep using fossil fuels? Can we leave such a world to our grandchildren?

WE HAVE TO THINK AHEAD

One of the problems of our society is that we have a short-term outlook. Our leaders and politicians want to get reelected, so naturally they plan to do things merely during the few years for which they are elected; they feel, quite reasonably, that if they do things for us that we like, we will re-elected, so naturally they plan to do things merely during greenhouse effect and the pollution of our atmosphere, we will need to plan on a much longer time scale; we need to think in 20-year, even 50-year, steps.

We certainly don't want the greenhouse effect to creep up on us, but every year that we do nothing the earth gets a fraction of a degree warmer. It is always tolerable to accept a small increase in temperature if we only think in one-year steps. But what about in a few years' time, when today's youngsters leave school? There are few nations that plan this far in advance, i.e. the time it takes for children to pass through the school system. Even fewer plan still further ahead. But we need to catch the greenhouse effect in time; we need to think ahead and plan what we will use to replace our present fuels that are responsible for a situation which is becoming impossible and dangerous.

However, many will tell you that if we give up our short-term way of thinking we immediately restrict our freedom to make choices in the future, and that threatens the principles of democracy. The key, though, is common sense and a reasonable attitude. If we make decisions that exploit the environment with no regard for the future, then not only is democracy threatened, but life itself. There won't be much freedom if the greenhouse effect is fully upon us.

5
THE FOLLIES OF ACID RAIN

When water vapour in clouds condenses into water droplets and/or snowflakes, they fall through the atmosphere mixing with the polluting gases such as the oxides of sulphur, nitrogen and carbon, forming dilute acids – sulphuric acid, nitric acid and carbonic acid. The first two are very strong corrosive acids and are cited as the main culprits causing acid rains. However the third one, carbonic acid, although much weaker than the first two, could actually be more damaging because it is produced in much greater quantities.

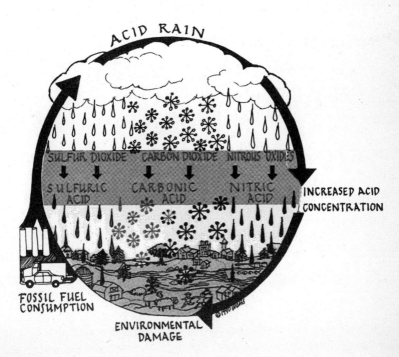

The rain containing these acids falls everywhere – into lakes, rivers and oceans, over forests, fields and farms, and on to homes, buildings and structures. Everything which comes in contact with rain water is subjected to the corrosive effects of the acids. These acids are harmful to everything, both to living beings and to material objects, and as the acid concentrations increase so the rain water becomes more destructive. And the acid concentrations increase as fossil fuel consumption increases.

EFFECTS ON AQUATIC LIFE

A complex interaction between acid rain, the water chemistry of the river, stream or ocean, the soil type, and the water and land use patterns all affect the overall acidification rates of an area of water. As the lake or estuary waters become more and more acidic – as a result of acid rains and acid snows – the

'Well, the rivers can't support salmon reproduction, so I decided to support it myself. I can't bear the thought of a world without lox.'

aquatic fauna and flora are affected. First the less tolerant organisms such as bass, salmon, trout and mussels will suffer; then those more tolerant to acids, such as water beetles and weeds, will disappear.

When the plants and/or animals in a lake or estuary die as a result of increased acidification, they start decomposing. Decomposing organic matter combines with oxygen in the water, resulting in a reduction of oxygen levels, and making the lake even less habitable for other life forms. The organic matter also enhances the growth of certain undesirable algae, which cover the lake's surface and reduce the penetration of oxygen and sunlight. This further reduces and destroys the life-supporting capacity of the lake or estuary.

Studies show that acid rains have already caused wide-spread acidification of many aquatic ecological systems (lakes, ponds, estuaries) in the northeastern United States, Canada, Norway, Sweden and the United Kingdom. For example, it has been reported that 4,000 lakes are fishless in Sweden, while a total of 14,000 lakes have been acidified. In the United States, more than 200 lakes in the elevated parts of the Adirondacks in New York State are dead due to acid rain deposits. This damage to aquatic ecosystems is evident, but it is not easy to estimate on an economic basis.

For example, several estimates of the ecological impacts have been reported for the Adirondack lakes. The estimates of physical effects, which are important determinants of eco-logical damage, are expressed as changes in fishable hectares and catch rate. About US$13 million worth of damage has been calculated on an estimated 10 per cent reduction in fishable areas. Extending this value to other regions of the United States where lakes have been acidified, the total loss in the United States for lake fishing becomes about US$1 billion per year. However, the above estimate is incomplete. It does not consider recreational value and the loss of tourist income for the regions in question, which could double this figure.

Another approach to estimating the damage to lakes because of acid rainfall is to calculate the cost to correct them. One technique which has been implemented to correct such acidification is to add lime to lake waters; lime is alkaline and thus can neutralise the acidity, permitting life to flourish again. The annual cost of liming was estimated to be US$40 million for Sweden in 1985. If lakes in the north-eastern United States were limed, the expected expenses would add up to some US$500 million per year. And of course, such remedial action can only be taken as long as limestone is available.

Acid rains don't fall only into lakes, ponds or estuaries. They also fall into rivers and oceans. In Canada and Scandinavia many rivers no longer support salmon reproduction. The acidity of the oceans is also increasing, such that it is estimated that by the middle of the next century the upper 60 metres (200 feet) of the oceans will be so acidic that they will no longer support aquatic life. No estimates of such losses exist yet.

DAMAGE TO FORESTS AND FARMS

Acid precipitations affect not only the waters of our planet, but the soil as well. In particular, the effect of acid rain on forests is quite destructive. As acid precipitation soaks into the soil, it dissolves ordinarily insoluble minerals and compounds, disturbing the organic nutrient balance in the soil. The dissolved salts, particularly those of aluminium, are toxic to the young roots of trees. Furthermore the acid rains also burn the tender new leaves of the trees, lowering the trees' resistance to disease.

The increasing damage to the world's forests is attributed to acid precipitation in general. Recently a comprehensive survey and analysis of forest decline was carried out in Europe. The results show that about 15 per cent of the total growing stock is severely damaged or dying, this damage being about six times greater than the annual felling of trees for lumber. Other studies show that in the Federal Republic of Germany, 50 per cent of the Black Forest is suffering damage from acid rains; the trees are sickening and dying daily as the air and soil become more and more inhospitable to them. There are reports indicating that the damage in Poland, Czechoslovakia and other central European countries is even greater because of their greater dependence on coal as a fuel.

In 1978 it was estimated that 5 per cent of forests in the United States were suffering from the effects of acid rains. This resulted in a loss of US$600 million for that year in terms of lost revenues. Add to this the loss of recreational and/or tourist income, which was estimated to be US$1 billion per year, and the total annual loss related to forests due to acid rains in the United States added up to US$1.6 billion for that year. But since 1978 oil and coal consumption has increased, so that the amount of acidic rainfall has increased. When one takes this into account, the projected damage to the forests in the United States due to tree loss and lost recreational revenues becomes some US$5 billion for 1990.

Not only do acid rains damage the plants in forests, but they also affect the quality and quantity of farm produce as well. Vegetables and fruits are smaller, more often deformed and less nutritious. They also reduce the yield by lowering the disease resistance of the plants. As a result, we may have shortages and end up paying higher prices for poorer quality produce. It has been estimated that in 1984 losses in farm produce in the United States due to acid rains were US$8.2 billion. When one considers the growing demand and inflation, then crop and produce losses due to acid rains rise to an estimated US$12 billion in 1990.

EFFECTS ON BUILDINGS AND STRUCTURES

When acid precipitation falls on our cities, the buildings and structures are adversely affected. Acid rain corrodes the stonework of buildings, including historic buildings. Such priceless treasures as the Parthenon in Greece, the Coliseum in Rome and Notre Dame Cathedral in Paris are among those that are showing the effects of acid rain. Germany is spending about US$4 million a year to replace the historic stonework of Cologne Cathedral; the work requires special stones and skilled stonemasons, and is tedious and time consuming. It takes a few years to replace the damage stonework on one side of the building, then the crew starts work on the next side, and so on; by the time they have completed the work on the last side of the building, the stonemason crew has to start all over again because the stonework on the first façade has already begun to corrode. Of course, many historic buildings are being totally lost because of acid rains – they can never be replaced.

Metallic structures such as bridges, railings and cables don't escape the effects of acid rain either, so the metals must be protected so that they don't corrode and fall apart. The same applies to buildings and other structures; they must be protected by frequent cleaning and applications of coats of plastics, chemicals and/or paint. Acid rain can even damage the protective coatings of paint and exterior metallic trims on cars, buses and other vehicles; hence, they must be cleaned and repainted more often to be kept in good condition.

Various estimates of the effect of acid rain on structures have been made. In 1984 it was estimated that the total cost to remedy the damage to historic buildings was US$20 billion worldwide. The damage to public buildings and other structures was estimated to be worth US$80 billion, and the expense to counteract the effects of acid rains on homes was estimated to be US$40 billion, adding up to US$140 billion per annum. This figure, if one takes account of the increased use of fossil fuels and the increased number of public buildings, structures and homes, becomes US$200 billion worldwide per annum in 1990.

EFFECT ON HUMANS AND ANIMALS

Acid rains increase the acidity of our drinking water. When this water flows through metallic piping it leaches out metals, such as lead, copper and aluminium, so that we end up drinking and cooking with a deadly soup of acids, poisonous metals and chemicals. In Sweden, for example, such acidified water was found to make babies ill.

But acidified drinking water affects adults as well as babies and children. Such water, containing toxic metals leached from watersheds and water piping, can cause various ailments, especially in the kidneys and urinary tract. Acidified waters lose their calcium content (we obtain calcium through drinking water in general), resulting in some cardiovascular diseases. It has been estimated that in the United States the detrimental effect of acid rains on human health in 1985 was, in money terms, some US\$180 million, corresponding to US\$220 million in 1990. This, of course, takes no account of the human suffering involved.

We can measure the toxic effects on ourselves of acid precipitation, and can try to clean up the waters we use, albeit at some expense, as pointed out above. However, this is not true for wild animals and plants; they just have to suffer the effects or change their habitats. But their habitats are becoming smaller and smaller as the human population grows and they also become less habitable as we continue to pollute the environment with the by-products of fossil fuels.

Hundreds of species of plants and animals become extinct every year due to the effects of fossil-fuel-generated pollutants and acid rains, but there are no estimates of such damage or losses in material terms. We have yet to understand the total impact of acid precipitation on our environment or, for

They just have to suffer the effects . . .

example, how the loss of one species of fish or grass can upset the balance of nature.

WHAT TO DO

Experts have predicted for years what would happen as our atmosphere is filled with the by-product gases of fossil fuel combustion, resulting in pollution and acid rains. But the human population at large is so buried in their 'instant gratification' view of life that looking forwards 20 to 30 years into the future is beyond their scope. They have immediate problems which need immediate solutions.

Well, the predictions made 10, 20, even 30 years ago, are starting to come true. Acid rains are now an immediate problem, and becoming more so. Our great-grandchildren may never have the opportunity to know what a wild deer or a forest looks like, except through pictures, and generations of people may never have the chance to see the Roman Coliseum. But we can reverse the process by, first, unburying ourselves from our 'instant gratification' view of life and begin to plan now for the future, so that we can hand on to our children and grandchildren a cleaner and healthier place to live.

6
THE HOLE IN THE OZONE LAYER

WHAT EXACTLY IS SKIN CANCER?

Everyone is familiar with moles and freckles. Moles are sometimes hereditary (as in the case of some beauty marks) and some are caused by exposure to the sun. Freckles usually appear in those areas most exposed to the sun, such as across the bridge of the nose and on the forearms. These are all very normal disruptions of the skin, and are harmless. Sometimes, though, what looks like a mole is really a form of skin cancer. Some cancers start out as a red patch of skin, while others start out the colour of the skin and later bleed and get a crusty surface. Still others, the life-threatening ones, start out as an irregular-looking mole with ragged edges, but unlike a normal mole, do not look exactly the same all over in their colour or shape.

What can make an otherwise normal looking mole into a cancerous one? Well, it's partly a matter of the cells of the skin itself, primarily the DNA (deoxyribonucleic acid), the constituent of each cell that determines the inherited 'blueprint' for the function and operation of that cell and the formation of other cells. Suppose your skin were exposed to a particularly large dose of ultraviolent radiation (the harmful rays) from the sun. This could upset the balance of the DNA in some skin cells, causing them to change their genetic blueprint. When this happens they will stop producing normal cells for your skin and start producing monster cells, cancer cells. That's when the blotches and tumours start to develop. Too much sun is dangerous, which is why you should protect yourself from the sun; be cautious and careful. If in doubt about how to do this, ask your doctor for the best method of sun protection for you.

WHAT IS SO IMPORTANT ABOUT THE OZONE LAYER?

Ozone is a molecule, three oxygen atoms joined together. Scientists therefore write it as O_3. (The oxygen we breathe consists of molecules of two oxygen atoms, and is written as O_2.)

Ozone occurs everywhere in the atmosphere, from ground level to 25 miles above the earth. At ground level it is an ingredient of smog so it's not a good thing, but in the outer atmosphere – the layer known as the stratosphere – it protects us from the damaging rays of the sun by absorbing and therefore screening out the ultraviolet rays that can cause skin cancers before they reach the earth.

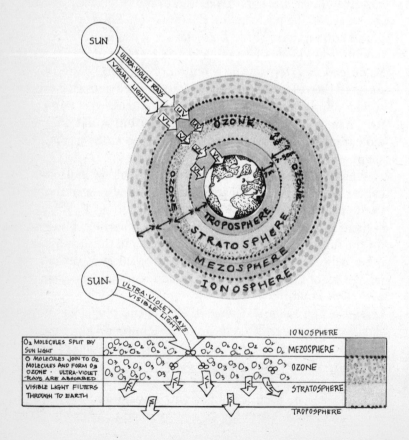

Once through the stratosphere, the sun's rays then have to pass through the oxygen, nitrogen, carbon dioxide, ground-level ozone and water vapour in the air, all filtering the sun's light before it actually reaches the ground and us. You might therefore think that hardly any sunlight at all would reach the earth's surface, but solar energy is very powerful. In fact, a great deal of solar energy escapes from the sun that we never see. The sun is so huge – 800,000 miles across the centre – that we only need one ten-millionth of the escaping solar energy to keep life going here on earth. And even that one ten-millionth of the sun's energy would be too much if it wasn't for the filtering effect of the atmosphere, and in particular of the ozone layer in the stratosphere.

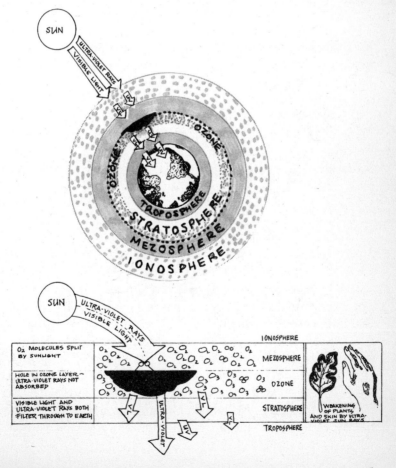

A HOLE IS DISCOVERED

In 1985 data from a NASA satellite orbiting high above the earth confirmed what some scientists had anticipated – a hole had formed in the protective ozone layer over Antarctica. Scientific teams from several countries were sent on Antarctic expeditions to take measurements of the ice and of the sun's damaging ultraviolet rays. And the research teams agreed – the hole in the ozone layer was definite.

This discovery did not bode well. Not a world catastrophe yet, because very few people live in the Antarctic, but suppose the hole were to spread? Other teams of scientists were called on to answer that question, physicians included. The overall consensus was that, for humans, a spread of the ozone hole meant a spread in the rate of skin cancer.

As you can well imagine, scientists and physicians alike have since been watching the ozone hole very closely. The disturbing news is that even though the hole in the ozone layer increases and decreases with the seasons, it is gradually getting larger. And skin cancer, just as predicted, has increased too. The last five years of the 1980s saw the greatest increase in the numbers of skin cancer victims than over all previous years. Some skin cancers are fatal if not treated promptly.

HOW DID THE HOLE FORM?

What has caused the hole must be man-made. The ozone layer must have existed for billions of years, otherwise the green plants wouldn't have developed as they have. And the fact that the hole originated, relatively speaking, in the last few years suggests that it is a consequence of our high-tech world that is causing the trouble.

Chemists know a lot about what substances react with what. As a substance there are only a certain number of things that would react with ozone strongly enough to 'eat' it. Suspicion soon fell on a man-made group of chemicals we know as Freons. Freons are also called chlorofluorocarbons – CFCs – because they are made up of chlorine, fluorine and carbon atoms. CFCs are used a lot in the manufacture of expanded polystyrene (styrofoam) and foam rubber (just think how many times a day you come into contact with these two materials), but most people know them as the stuff in their refrigerators and air-conditioners that keeps the air cold. They used to be used as a propellant in aerosol can products like hairsprays and room fresheners, but more and more countries are banning their use in this form.

CFCs escape from leaky air-conditioners and refrigerators, or out of a discarded expanded polystyrene (styrofoam) cup, and make their way up to the outer atmosphere where the ozone layer is. Once there they go about 'eating' ozone molecules; one CFC molecule can gobble up 100,000 ozone molecules over its 150-year lifetime and one expanded polystyrene cup contains one billion billion CFC molecules.

Why the ozone hole appeared over Anarctica first is not readily understood. Some believe that the CFCs were concentrated there due to the upper atmospheric air patterns and the spinning of the earth on its axis. Others believe it has something to do with the temperature at the South Pole. It may be a combination of all of these, as there is new evidence that the ozone layer at the North Pole is also starting to thin.

One CFC molecule ∈ eats 100,000 ozone molecules and lives for 150 years.

Originally, scientists believed that the ozone layer would thin over the whole globe, not in any particular spot, at the rate of about 2 per cent over ten years – not an amount that would cause alarm. But now alarm bells are ringing. In addition to the size of the hole in the ozone layer, scientists are worried because there is no clear answer to why it is forming where it is.

IS ANYTHING BEING DONE TO STOP IT?

The answer is yes, but is it enough? Probably not.

The first big step was taken at an international meeting in Canada. Many countries around the world signed a document, now referred to as the Montreal Protocol, in which they all agreed to cut CFC production by 50 per cent by the year 2000. But many environmentalists say CFCs have to be cut by 75–85 per cent in order to help the environment realistically.

One offer to help the ozone layer came from the US chemical giant DuPont, the inventor of Freons. DuPont announced it may have a substitute for CFCs that will not harm the ozone layer, but that it won't be ready to manufacture the new product until all the testing and analysis is done, which may be a few years to come. In the meantime, other chemical companies are racing to find a replacement chemical that is ozone friendly.

However, there may be a better way than chemicals. CFCs aren't the only way to keep refrigerators and air-conditioners cool. There is a technology that has been around for a long time, called thermocouple cooling. It works like this. If you pass electricity through specially designed wires, one end of the wire gets cool and the other gets warm. For refrigerators and air-conditioners, the cool end could be put on the inside where the air needs to be cool, and the warm end could be left on the outside. With this, not only would refrigerators and air-conditioners be quieter (no compressor noise), but there would be no CFCs escaping into the atmosphere and widening the hole in the ozone layer.

But perhaps the biggest thing to be done to stop the ozone hole has yet to be tackled – and this is something you could do. Educate the people around you about what a serious problem this is. Tell your friends what you have learned from this book. If you don't like the way your environment is being handled, tell someone. Get a group together and speak in a collective voice – the bigger the group, the more clout you'll have. Suggest ways in which the current laws can be changed.

Remember, time is an important factor here. It may take years to develop a substitute for CFCs that does not simply trade one environmental disaster for another. In the meantime, we have to stop using CFCs and all things made with them, and at the moment that is not just going to happen. So we have to make it happen. In spite of what we might think, in general we live in a world where government is by the people for the people. Let's use the power we all have.

7
SMOG, DIRT AND CANCER

When we leave a town or city and enter the countryside, we find an environment that is much cleaner, where much of the pollution that urban dwellers take for granted is absent. And when we return to or visit the city, the town or any other area where the automobile is frequently encountered, everything is dirtier. The air tends to be a little hazy and certain smells are noticeable. But we must be fair to the automobile though – it isn't the only cause of pollution in cities; there are some cities where the smoke and ash from chimneys add dirt to the air, all of which becomes part of the haze.

For example, have you ever noticed how, in the country, your nails remain clean, while in the city you are always having to clean them? Where does this extra city dirt come from?

SMOG – WHAT IS IT?

Everyone is familiar with mists. Mists are nothing but suspended tiny water particles; they are a dilute white cloud, and are quite clean. You can walk through a mist feeling a bit clammy as a result, but your eyes do not smart, you don't cough and you don't feel grimy.

Fogs are thicker, making visibility difficult, but are not much different than mists, for they too are basically water vapour. The one distinction is that, in a fog, there are solid particles present on which the water vapour has condensed; the presence of these solid particles is why fog is thicker than mists.

Smog on the other hand, is completely different. Smog is a modern phenomenon. First of all, the appearance of smog is more threatening than the appearance of fog. A very dilute smog, in which visibility is still several miles, will cause a gritty irritation of the eyes and a nasty grittiness in the throat. As the smog gets denser the effects increase, the bronchial tubes become irritated and coughing begins. The last stage of the smog is the reduction of visibility to such an extent that driving becomes difficult or, finally, impossible, although this only occurs occasionally now.

The vital difference between fog and smog lies in a new component, a complex organic compound condensed on minute pieces of dirt – often on pieces of fly ash or evaporate from a furnace fire, or simply on dust floating in the air. And what is this organic compound that causes so much fuss? It is called peroxyacyl nitrate (PAN), and it gets into the eyes and throat, causing burning and irritation if you breathe in too much of it.

WHAT CAUSES SMOG?

The way smog is formed is quite complicated. We won't go into every grimy detail here, but we will give you an outline of its causes and effects.

The story of smog starts with the automobile and with the materials that come out of its exhaust pipe. In particular, two of these chemicals are relevant.

The first is a simple compound called nitrogen dioxide (or nitrous oxide). When this comes out of the exhaust of a car, and if there is plenty of sunshine, then it dissociates and forms nitrogen monoxide (or nitric oxide) and another compound called 'active oxygen'. The active oxygen atoms easily combine with the regular oxygen atoms (the kind we breathe) already in the air, forming a new compound called ozone. This ozone is harmful because it remains at the ground level, not like the protective layer in the stratosphere you read about in Chapter 6.

But let's leave ozone where it is for a second, and come back to the other materials that come out of the vehicle exhaust. When petrol burns in an internal combustion engine, it forms some carbon-containing compounds. The molecules are quite long and snaky in appearance, but are far too small to be seen with the naked eye. More importantly, they are remarkably chemically active; they float around in the air and combine with the ozone that has just formed, as described above, producing the PAN we have mentioned already.

This subsequent PAN molecule is rather a fearsome-looking compound – large as molecules go – containing 'active oxygen' and nitrogen oxides too. It is this compound that is adsorbed on to the small particles of solids present in the air. These tiny particles and the PAN molecules absorbed on top of them, floating in the air, are the essential constituents of smog.

WHAT SMOG DOES TO YOU

We have stated above some of the symptoms of 'smogitis' – the eyes smart and then, as the smog increases, the bronchial tubes are attacked and become irritated. If someone is already ill with a respiratory disease, and if the smog is thick enough, death may result. Other ills arise from repeated exposure to smog, a series of ills that are sometimes referred to by the catch-all phrase of 'smogitis'.

Sometimes, though, it gets much worse than we have described, such that from time to time we get a killer smog. The most well-known smog of this type in history was that which attacked London in 1952. It lasted about four days, during which time it was difficult to walk in the streets because you could hardly distinguish the pavement from the road. It invaded houses and made them so gloomy that you could hardly see to climb the stairs. And 4,000 people died due to health problems associated with the smog.

SMOG AND THE AUTOMOBILE

Those who have read this chapter so far will understand well that smog is a product of the fuels we use – petrol and diesel – in our present forms of road vehicles. It is essential to have such transport; indeed personal transport has become the essence of late 20th century life in the developed world. But it is not necessary to run these vehicles on petrol or diesel oil.

There are other cleaner ways, and later on in this book we will tell you what they are.

CANCER FROM DIESEL AND PETROL

We have learned that the use of diesel and petrol to run our cars, and oil to run our industries, has given us a wide range of problems to cope with – the greenhouse effect and the heating up of the planet; general diminished good health especially respiratory diseases such as emphysema; acid rain; deforestation and the death of lakes. However we do not generally associate petrol and oil with that most dreaded of diseases – the development of cancerous growths and tumours. Recent research has established a correlation between compounds found in exhaust fumes from diesel driven cars and certain types of cancer. For example, one study by the American Cancer Society found correlations between non-smoking lung cancer deaths and the number of vehicle miles travelled by those individuals. Diesel engine exhausts are especially carcinogenic because they contain a compound called benzopyrene, a chemical known to cause lung cancer in cigarette smokers.

But this is not all. In areas where petrol and oil are produced and refined, there is an increased incidence of cancer compared to areas where no such industry exists.

It is very sad that, despite all the many advantages of the automobile, the cost to human health, not to mention the environment, is so great.

GIVE UP THE AUTOMOBILE?

But it would hardly be practical to give up the automobile. It is part of our lives, much as horses were for earlier generations.

Some people have suggested that methanol could be used as a substitute for petrol and diesel; they say it does not cause the same pollution as petrol (for example, no smog), which is true enough. But what they don't tell you is that it emits as much carbon dioxide as petrol, as well as producing yet another carcinogen, formaldehyde. So would we be any better off? Replacing oil and petrol with methanol would be like jumping out of the frying pan into the fire. Instead we need to use a fuel that is clean and efficient that does not cause cancer, smog or dirt. Hydrogen is such a fuel, and we shall explain why later in this book.

We are not asking anyone to make unreasonable sacrifices. We are merely asking people to understand the decisions that have been and are being made for them. We are asking people to realise that the selfish attitude we have all adopted and the demand for development at all costs is costing us our health and the environment.

Give up the Automobile?

DIRT

But now we want to tell you a little about dirt. We don't just mean household dirt, but the dirt you find everywhere around you: the dirt that collects in open places after they have been cleaned; the oily grime that you can scrape off most smooth surfaces – smooth white painted shelves, for example – and which means that to keep something clean you have to be constantly at it. Where does all this dirt come from?

Well, it comes from the electrically charged dirt in the air. All particles that stay up in the air for a significant time become electrically charged; this keeps them apart, preventing them from sticking together and becoming heavy enough to fall to the ground. But eventually their charge is reduced or they lose it, and they stick together into bigger particles and settle down on the surface of the earth.

A number of these particles come from natural sources and were here before the automobile. They come from forest fires and from the enormous amount of ash poured into the air from the eruptions of volcanoes that are occuring more or less constantly all over the planet.

But this century we have put a great deal more of such particles into the world by burning the fossil fuels – petrol, coal, oil, wood – we use to propel our cars, to generate our electricity, for heat and to cook with. Diesel oil in particular produces an exhaust that is over-rich in small black particles. So our present fuels don't only produce smog, but also the dirt and grime that causes us so much extra work.

8
THE LESSONS OF NUCLEAR ENERGY

The birth of nuclear energy was the explosion of an atomic weapon in Alamagordo, New Mexico, in 1942. World War II was subsequently ended by the dropping of the atom bombs on Hiroshima and Nagasaki – the Japanese surrendered when faced with this devastating weapon. Shortly afterwards it was announced that there would be an 'Atoms for Peace' programme, and governments began to research and control nuclear power, with the expectation that within 50 years entire national economies would be driven very cheaply by electricity from nuclear power.

Nuclear fuels and chemical fuels – petrol, jet fuel and natural gas for example – differ tremendously in the amount of energy we're able to get from them. Furthermore, nuclear fuels don't necessarily have to have another substance to react with in the way that chemical fuels need oxygen in order to burn; nuclear fuels just disintegrate spontaneously, and if this happens at a 'reasonable rate' there is a steady production of heat, the heat is used to produce steam, the steam drives turbines which work generators, producing electricity.

HOPES AND FEARS

At first, nuclear fuel seemed to have huge advantages over the chemical fuels; during the late 1940s and early 1950s there was hardly any argument but that we should move to nuclear fuel as soon as possible. Many thought the Garden of Eden had arrived; there were pictures, for example, of a thimbleful of nuclear fuel showing the quantity required to run a town for weeks. And nuclear power was going to be almost negligible in cost – a few tenths of a penny per kilowatt hour. The point was often made that at that price, there would be no point any more in turning off lights to save energy.

With all we know now about the effects of carbon-containing fuels, nuclear power, which does not produce carbon dioxide, might seem to represent the answer to all our present environmental problems. But there are important drawbacks with nuclear power. For a start nuclear energy has always been associated with its violent use as a gigantic weapon of war, which is one of the reasons why there is still a fear of nuclear energy. We fear that, no matter how peaceably we exploit the awesome energies produced by a nuclear reaction, it could get out of hand and create an unintended explosion. In fact, such a possibility is very unlikely. Not even the nuclear accident of 1988 in Chernobyl involved an atomic explosion, the kind that occurs in an atomic bomb, despite the nuclear fire which burned for many days. But there are considerable dangers from nuclear reactors, and even though the matter has become clouded with controversy, a few points are very clear.

RADIATION

Now, we always live with a tiny bit of nuclear radiation. This comes from the earth, and is called background radiation; a good example of naturally occurring low-intensity radiation is the small quantity of radon gas that leaks up into many houses. We must assume that such low-intensity radiation is harmless, because we have developed and evolved over billions of years under the influence of this natural radiation.

However, if the human body is subjected to high-intensity radiation from radioactive particles – the kind of particles in the fuel of a nuclear reactor – there is no doubt they will harm the body, in two ways.

- There is the type of harm that is bullet-like; the particles crash into the cells and destroy them. The result is often a tumour.

- The other kind of harm is indirect. The particle itself gets into the water-like fluid that surrounds the cells in our body. When this happens, a special kind of harmful oxygen (ionised oxygen) is produced that is not normally present in the cells. The cells then soak up this poisonous fluid causing their growth patterns to go awry. Again, this can cause cancer.

About the only fact in this area that is not clouded by controversy is that too much nuclear radiation causes cancer. But how much radiation is too much, and how much really escapes from nuclear reactors? How much affects people in the vicinity of a nuclear reactor, or causes harmful reactions to take place in stored food? How long after the exposure to radiation will it be before cancer develops? You will get different answers to these questions, depending on whom you ask.

It seems indisputable that properly built and properly maintained nuclear reactors do not give out significant amounts of high energy radiation; the builders of such plants are probably correct in saying that a properly constructed well-managed nuclear plant is harmless. However, we know from experience that the plants aren't always well managed, they do have defects, often due to human failings – and that's where the trouble begins; accidents can happen and, if fire breaks out, as it did at Chernobyl, the smoke that is produced contains nuclear particles which can travel long distances. This smoke is then extremely hazardous to anyone who comes into contact with it.

An important point concerning the health hazards from nuclear radiation is the intensity of the radiation. This means how much radiation a person receives over a certain period of time, and in low-intensity radiation the amount over time is the key. Thanks to a Canadian physicist named Petkau, we now know that if radiation particles are ingested slowly, they have a worse effect on the human body than if they are ingested quickly.

It's quite easy to understand how this occurs. Remember how high-intensity nuclear particles can produce harmful ionised oxygen in the liquid around the cells? The same thing occurs in low-intensity radiation. The ionised oxygen in the fluid around the cell gets on to the cell's surface and can affect the DNA, the genetic blueprint that controls the structure and function of each cell. Damage to the DNA can therefore upset the functioning of the cell and can even result in birth defects if cells in the reproductive organs are affected. When radiation enters our bodies quickly, as in the case of high-intensity radiation, many ionised oxygen particles are produced at one time, and they cannot all be absorbed by the cell. Many are therefore 'wasted'; they can do no damage. But if we are exposed to the same dose slowly, over a long period of time, every ionised oxygen molecule has a chance to get on to the surface of the cell and affect the DNA.

Let's end the controversy over the health hazards of nuclear radiation contamination by saying that, if all goes well at nuclear plants, there is little danger. But it seems to be a fact of life that accidents do occur and when they do, high-intensity radiation, direct and cancer-causing, will be a health threat. Furthermore, low-intensity radiation, if received over a long time, may do much more damage than originally thought.

NUCLEAR CATASTROPHES

The one thing that the public fears more than the subtleties of low and high-intensity nuclear radiation is that the whole nuclear plant will blow up – an atom bomb near a centre of population. The truth is, such an occurrence is not very likely. A melt-down is a more realistic fear than a nuclear explosion.

In a melt-down the normal cooling water that keeps the reactor in the plant going is accidentally interrupted. The reactor then overheats and the solid part of the core melts. Water now leaks on to the reactor from breaks in the cooling water lines, and steam at very high pressure is produced, probably blowing the top off the reactor. Some of the contents of the interior of the reactor, including the nuclear fuel, would then be thrown into the surrounding atmosphere and carried for great distances. In addition, the very hot radioactive core would simply melt it's way down through the reactor housing to end up about a mile deep in the earth.

A disaster of this kind has been foreseen, and planned against, by the designers of the nuclear power plants; the hope is that those shiny domes you see over most reactors would contain the explosion caused by overheating. However the planners cannot be certain, and no one wants to experiment. Fortunately, we have not yet had a true melt-down, for if we had one near a town it could cause the eventual deaths of hundreds of thousands of people. But we have had some near misses, and there is a big lesson to be learned from all these near misses; they all involved human error.

In one such near miss, at Brown's Ferry, Alabama, a careless technician passed through some underground passages holding a candle. He had the candle in his hand because he wanted to see which way the air was moving in the passageway. Getting tired of holding the candle in front of him, he held it above his head. Some of the insulation on the pipes containing the cooling water caught fire. Soon the cooling mechanism was interrupted and a nuclear melt-down was on its way. The reactor and the town (maybe much of the state) was saved by emergency action.

In the more recent disaster at Chernobyl the essential cause of the accident was that the scientific staff began experimenting with the reactor while it was in full operation. They wanted to see what would happen if they turned down the spinning rotors – the turbines – that are used to make electricity. In order to carry out their experiment, they turned off the automated safety precaution circuits that protected the reactor. While they were carrying out the experiment, heat production in the reactor got out of hand, and there was no automatic mechanism operating to shut it off. A nuclear fire started and, because the reactor had no 'shiny dome' over it, smoke containing 'hot' nuclear material spread over much of central and northern Europe.

In both these accidents there was no fatigue of materials, no wearing out of metal parts, merely human error. And the fact that people can be careless is one of the main lessons we have learned in dealing with nuclear reactors.

NOBODY DIED ...

Some experts say that many thousands of people will eventually die as a result of the catastrophe at Chernobyl, even though the official death toll was only four. There is obviously a great discrepancy between these two figures and this is characteristic of differences in opinion as to who gets hurt by nuclear reactors.

The principal health risk to people from leaks of nuclear radiation is the formation of cancerous tumours and leukaemias. But these diseases will not develop for many years after the initial exposure to radiation, by which time it can be difficult to prove that nuclear radiation caused a particular cancer. For instance, by the time the problem is detected, the victim may have lived in many different parts of the world and been exposed to many hazards.

Nevertheless, by experimenting with animals, it is possible to predict how much cancer will eventually develop from exposure to a certain amount of radiation. On this basis, health officials predict the deaths of 20,000 people as the final result of the catastrophe at Chernobyl.

It is even possible to calculate the chance of a nuclear accident. For example, were the United States run entirely on nuclear power, about 2,000 nuclear reactors would be needed. If we apply the known accident statistics to these reactors, an accident would occur about once every two months.

THE IDEAL – NUCLEAR FUSION

So far we have been discussing how we can obtain immense amounts of energy from very tiny amounts of uranium, harnessing the energy that is set free when the nucleus, the central dense part of the atom, splits open. Such a reaction is called fission – splitting apart.

But there is another way of obtaining enormous quantities of energy, an idea not as old as nuclear fission. Indeed, it is the opposite of fission; instead it involves nuclear fusion – the combining together of atoms.

If two hydrogen atoms are fused together under extremely

high pressure and temperature they form another element, helium, not more hydrogen, while at the same time releasing an immense amount of energy. The fusion of hydrogen atoms to helium is the source of the sun's energy.

Nuclear scientists have always maintained that, if fusion could be achieved under controlled conditions, it would be the ideal way to get useful safe heat from nuclear energy. But there is one hitch. How do we get the hydrogen atoms to fuse in an orderly controlled manner, in a power plant, so that the released energy can be utilised to produce electricity?

We know nuclear fusion works, for we have the sun and the hydrogen bomb as examples. But the bomb, though producing lots of energy, does not produce it in an orderly controlled manner. So, until we are able to harness this energy safely we will not be able to tap this colossal source of power and produce electricity from it.

NO TO NUCLEAR POWER

For a variety of reasons the construction of nuclear power stations around the world has slowed dramatically.

In the US this has been achieved partly by the use of court injunctions – temporary restraining orders issued by judges. If a properly reasoned case is made by experts in environmental safety, backed up by the advice of lawyers, the judge may be persuaded to issue an injunction against a public utility, whereupon they have to stop work on the nuclear power station immediately. Thereafter, the allegations made by the lawyers and environmental experts have to be tested in a courtroom battle of expert witnesses. If these arguments hold up, the court will then issue a permanent ban on building a reactor at that location. People in many areas of the US have used this procedure to block, or at least slow down considerably, the construction of nuclear power stations. Furthermore, because such a procedure can be expensive, it has raised the overall costs of building such plants.

In the UK the commencement of any such major civil engineering project is invariably preceded by a public enquiry, at which those for and against the project elaborate their arguments in front of an impartial adjudicator. Again, this is a lengthy and expensive procedure.

However, the principal reason that there has been a slow-down in the construction of nuclear reactors is far more simple – the electricity they produce is very expensive. They are extremely complex to build, and as each year passes ever more safety measures have to be included in their design and operation.

Furthermore, once a reactor gets to the end of its useful life – and they don't go on for ever – it has to be decommissioned. This means removing all the radioactive materials that can be dealt with safely and sealing off the remaining structure in a concrete jacket. Again, this is a very expensive procedure, and adds even more to the overall cost of producing the electricity. Additionally, the price of uranium is no longer as low as it used to be.

All these factors taken together mean that the idea of nuclear power producing an inexhaustible supply of very cheap electricity has become a myth, and this has slowed down enormously the worldwide construction of new nuclear power plants, while in some countries the nuclear power programme has ground to a complete halt.

WHAT HAVE WE LEARNED?

So, the nuclear age seems to be passing. People don't want nuclear energy, and they have made that pretty clear. There is no doubt that if the people are conscious of a danger, they will find the legal means to stop whatever the danger is.

We have learned that human failing causes nuclear accidents. There seems no way to eliminate people completely from the operation of the nuclear plants, or to guarantee that they will perform with 100 per cent diligence 100 per cent of the time, so we are subject to the weaknesses arising from human nature.

People have to think many decades ahead. If we had thought decades ahead about the nuclear energy business – about safety and decommissioning, for example – we wouldn't have built nuclear reactors in the first place. But we did not foresee numerous things – the tremendous expense of construction and maintenance, the rapid exhaustion of the fuel, the health hazards caused by using nuclear power, and the fear and anger of the people.

We have learned that so-called common sense does not necessarily always hold good when assessing nuclear dangers. For example, who would have thought that, for identical doses, low-intensity radiation was more damaging than high-intensity radiation?

We have learned about safety, though; because nuclear energy was so new, we learned how to handle it as we went along. For instance, most safety analyses are concerned with whether or not pipes, cooling systems and valves will last and remain reliable. But the nuclear accidents have always occurred because of unpredictable human behaviour. We have therefore learned that we need to guard against human unpredictability better than we have in the past.

Because we were all overly optimistic of nuclear energy, we moved ahead very quickly – in retrospect, too quickly. You should be aware by now that we have to change our energy system. While we develop these new energy systems, let us not forget what we have learned from nuclear energy. Let us remember to tell those who guide our policymakers to point them in the direction of energy that is clean, renewable and safe.

9
A RECAP

In the preceding chapters, we have presented the major threats to our environment. Unfortunately, no one knows *precisely* what is happening to the environment, nor *exactly* what it all means. However, one thing should be clear by now and that is carbon-containing fuels are damaging the world we live in – and that affects you and your children.

WHAT CAN ALL THIS DO TO YOU?

- If we have to cut back on the use of energy – because using energy generated by fossil fuels is what is causing all the trouble – a major impact would be increased unemployment especially if we don't begin educating and training people in the new energy technologies of the future.
- Then there is the loss of land we shall suffer as the seas rise. It may seem that a rise in sea level of a few feet in 50 years is not a great threat, but a few feet is the conservative side of the estimate. And even a few feet can have a devastating effect worldwide. But some scientists say that eventually the rise may be as much as 24 feet (7 metres). The amount of land lost as the seas rise would then be enormous. Where do *you* live?
- In a couple of generations it may simply become too hot and arid to sustain wheatfields in the large areas of the world where wheat is currently grown. Will your grand-children have enough to eat? Life expectancy would also fall as nitric acid and smog in the atmosphere reduce our resistance and thus encourage the spread of disease.
- Another potential disaster that looms unclearly ahead is what happens when deforestation takes hold? This will affect the air that you breathe. At present there are many sources of carbon dioxide in the atmosphere; from the exhaust pipes of cars, from power stations and factories, but also from the air we exhale – humans and all the animals of the earth breathe out more CO_2 than they breathe in. In the past carbon dioxide levels were kept in check by the photosynthesis of green plants – this process needs CO_2. With the vast reduction in the areas of trees that cover the earth, and as the population grows, the exchange of CO_2 gets out of balance.

IS THERE A REMEDY?

Some alternatives and remedies have been proposed, and the most popular solution being presented at the moment is that of using methanol as a fuel instead of oil. However, as methanol also contains carbon, this substitution, despite having the limited advantage of not contributing to smog formation, would produce virtually as much carbon dioxide as would be produced from burning oil. Another idea is to remove the carbon, released by burning fossil fuels, from the atmosphere. It is possible now to filter the air through certain chemical compounds and to remove the carbon dioxide and other toxic substances in the air. It has even been suggested that the filtered carbon be converted to methane (natural gas), which, when burned, would release CO_2 back into the atmosphere, but as the CO_2 would have been collected from the atmosphere in the first place, there would be no net increase. Another future possibility is that in 60-100 years' time it may be economically more viable to take carbon out of the atmosphere – for instance, for textile manufacture etc. – than to derive it from coal, which again would reduce the amount of CO_2 being released into the atmosphere.

Of course there have been many attempts by advocates of the carbon-containing fuels and those who back them – industrial magnates who make the profits – to assert that more study of the effects of these fuels is needed before remedial action can be taken. Obviously more studies are a good idea as we need as much information about pollution as possible; but there is no reason to stall on the initiation of change. We already know for certain that fossil fuels pollute our atmosphere. There are many details of *how* this happens that we do not know about, but we are certain that the damage is caused by using carbon-containing fuels, and we must eliminate them completely as a source of energy. This is the principal and clearest message of this book.

Another message is that the problem of carbon-containing fuels can be dealt with completely by our governments taking appropriate action, and making them do this is up to you.

THE REAL REMEDY

So, the real remedy lies in the hands of the people – you. Indeed, this book has been written to give the reader the ammunition she/he needs for a people's assault on the corporate and political establishments to persuade them to stop the use of the carbon-containing fuels.

There is only one solution to the greenhouse effect and all the other problems we have outlined, and that has to be faced by the manufacturers and distributors of our present carbon-containing fuels. We have to eliminate the carbon from the fuels we use today, and instead merely make use of the essential element, the hydrogen. The use of pure hydrogen as a fuel would produce no CO_2 whatsoever; in fact, apart from tiny amounts of NOX if used as a fuel for engines, nothing harmful is produced from hydrogen combustion because the main by-product is water vapour. Thus the use of hydrogen as a carbon-free fuel would not only help to stop the greenhouse effect, but would also eliminate acid rain and other pollutants – needless to say, we would all be happy to see the removal of the carcinogenic properties of some of the constituents of the present carbon-containing fuels.

Replacing our present fuels with hydrogen is the only energy scenario that can give us a high standard of living without the greenhouse effect, without dangerous pollution and without the threat of nuclear disasters such as Chernobyl. Perhaps one day the fusion of atoms will be able to contribute too to our energy needs, but at the moment such a prospect is probably more than 50 years away, and good prospects for the economic production of electricity by nuclear fusion perhaps further still.

So, after uncovering the facts in this book, it will seem as remarkable to you as it does to us, that the main objective – the need to eliminate carbon-containing fuels – is not the heart and soul of all policies for combating the greenhouse effect, and of every policy for the future of the environment. But perhaps you can see who would lose politically and economically if fossil fuel production were taxed into decline.

PART TWO

10
SOLAR ENERGY – THE BEST ANSWER

Most people are familiar with rooftop solar panels. They have a black background so they absorb the sun's heat, and then water is run over the hot surface; this hot water can then be used, usually to heat up the domestic hot water. This is often a satisfactory method of supplying kitchen and bathroom water, the solar panels paying for themselves in terms of reduced electricity bills after three to ten years.

But the sun can also make electricity. The discovery that sunlight could be converted to electricity was made by Pearson, Chapman and Fuller of the Bell Telephone Laboratory in 1954; they shone sunlight upon two different pieces of silicon, joined together like the layers of a sandwich. A circuit developed, because of the way the silicon for the sandwich was made, and once this had occurred electrons could pass easily, creating an electric current. This electricity could then be used to power a light bulb, an electric motor or, if enough silicon sandwiches are joined together, a town. As an example, one square metre of silicon sandwiched together would give us 100 watts of power (about a twentieth of the electricity needed to power a house), as long as the sun was overhead and there were no clouds. The creation of electricity from sunlight is of immense importance, and could be a future source of world energy.

WHY PEOPLE USED TO BE AGAINST SOLAR ENERGY

For many years there was an impression that solar energy would be impractical because it was thought to arrive on the planet too diffuse, too dilute, to be used. Another reason people were against it was because of the high capital costs involved. The specifically designed silicon sandwiches are a form of photovoltaic cell, and in the past photovoltaic cells had to be made by hand; each layer of the sandwich needed someone to grow silicon crystals, one by one, until there were enough to make a layer. It was far far too expensive to make enough silicon sandwiches for large scale use, so many people thought solar electricity would always be too costly.

However, new discoveries in photovoltaic cells have changed the situation. Now there is a new method of making photovoltaic cells that does not involve the painstaking process of growing crystals one at a time. A new material called amorphous silicon (silicon with no ordered crystal structure) is involved, and can give excellent yields of electricity from silicon sandwich cells, although not as much as the expensive single crystals. It is this invention of amorphous silicon that has made a revolutionary difference to solar electricity, a difference that may lead us towards a solar energy-dominated world.

Another concern is the amount of land that will be needed for 'solar farms'. Solar farms are tracts of land with arrays of photovoltaic panels similar to the ones you see on many rooftops; the difference is that these arrays go on for miles. The usual argument against such solar farms is that land that could be used for agriculture would instead be covered with solar panels, and would thus be useless for growing food. Well, it is true that the area required to collect enough light to run a city of half a million people, say, would be quite large – about 60 square miles ($155 km^2$). However, this land would not be prime agricultural land, but arid desert.

For example, if we assume that all the solar energy collection of the whole world could be sited in the desert areas (which, of course, wouldn't happen because some home-

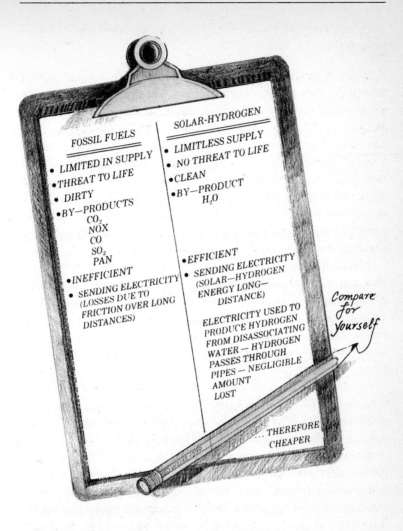

owners would have their own solar collectors, just as they do now), then it would take up only a small fraction of desert land (about 10 per cent of the world's deserts). So all the solar energy collection of the world could take place in areas that are pretty well unusable for any other purpose.

Conservationists might voice concerns about disturbing desert environments as they are an important habitat for many species of flora and fauna. Obviously some disturbance of the deserts would be inevitable when the equipment for solar energy collection is installed, but there would not be any long-term disruption or damage.

How the solar energy would be collected

We envisage a series of heliostats which would be supported about 10 to 20 feet above the ground. Large mirrors, perhaps 20 feet square, would be oriented towards the sun's path across the sky. The sun's rays would be reflected by the mirrors and directed towards a bank of photovoltaic cells. One effect that these structures would have on desert life would be the shadows they would cast which might cause a fall in temperature of 10°C (50°F) as the surface of the desert loses heat quickly once out of the sun. This change in temperature would most likely affect the insect population. However, as desert fauna they are adapted to big differences in temperature between day and night time. Also, the mirrors wouldn't cast shadow over the entire area of the solar farm as space would be needed between them for maintenance vehicles.

LIGHTING THE NIGHT

Another big misunderstanding about solar energy is that it has never been clear to people how it could possibly be used at night. Lighting cities at night with energy from the sun seems far-fetched at first, but we have to remember that if we are going to collect energy in far-off deserts, then it is important to be able to transfer solar energy from these places to population centres, which could be a thousand miles away. Such long-distance transmission of energy via electricity in cables is inefficient because so much of it is lost along the way because of the resistance of the wires.

To get around this problem we would have to convert the solar energy into some sort of energy store that could go through pipelines or be transported in containers, and the simplest option would be to use the solar-generated electricity to split water into hydrogen and oxygen, using a device called an electrolyser. This hydrogen could then be pumped through pipes at high pressure. When it gets to population centres the hydrogen could then be used directly as a fuel and/or be put through a device called a fuel cell, a reverse electrolyser, which would convert the hydrogen back into electricity. (Fuel cells have been used successfully aboard NASA space missions to produce electricity and water for the astronauts; and a large fuel cell has been operating in Tokyo, Japan, producing some of that city's electricity since 1984.)

Because of the revolution that has occurred in the manufacture of the photovoltaic cells necessary to produce hydrogen from water, Professors Ogden and Williams from Princeton University have shown that economically competitive hydrogen can be produced (via the use of solar panels made from amorphous silicon) not in some distant part of the next century but by the end of this century, if only we have the will to develop the technology fully.

ABSOLUTELY NO POLLUTION

The solar energy scheme which we are putting forward here is sometimes called the solar-hydrogen energy system. Put simply, solar energy is converted to electricity; to get the electricity over long distances, or for use at night, the electricity is used to electrolyse water to produce hydrogen (only the electricity that is not needed immediately by businesses, factories and households is used in this process). The hydrogen is then sent through pipelines, just as natural gas is sent today, to cities and towns.

The benefit of this system is that by putting excess electricity (the electricity not immediately used) to work by producing hydrogen, it is not wasted. In addition, it is cheaper and more efficient to transport hydrogen through pipelines than to send the excess electricity over wires and through cables. Finally, and perhaps most beneficial of all, hydrogen and solar energy do not pollute; when hydrogen is used to supply heat or energy, water is the by-product. Such systems produce none of the dreaded CO_2, no sulphur to cause acid rain, and no harmful pollutants to cause smog. Our solar energy will be around for a few billion more years – it is everlasting as far as we are concerned – and we would obtain hydrogen from

water, and that won't run out either as water is the by-product
of burning hydrogen. Thus solar-hydrogen energy is a clean
and renewable system.

So many of the natural processes on our planet are self-
sustaining – the circulatory systems of animals and humans,
the respiratory system of animals, humans and plants, the food
chain, and the earth's water cycle. Doesn't it make sense that
our energy should be derived from a renewable system too?

II
SOLAR-
HYDROGEN
COUPLING

ADVANTAGES OF SOLAR ENERGY

In this chapter we will look in more detail at the energy system we proposed in the previous chapter.

Life on this planet would not exist without solar energy (sunlight). Our lives wouldn't be as comfortable or as convenient without electricity, oil and petrol, but we could stay alive without them; without the sun, however, all plants and animals would die, and life on earth would end.

When the sun – the power that keeps us all alive – is used to make electricity, it is a clean and safe form of energy. Furthermore, solar energy is almost inexhaustible. We say 'almost' because, in order for the sun to give light and energy to the earth and all the other planets in our solar system, it is slowly burning itself up; but don't be alarmed, as scientists estimate that the sun will last a few billion years more before it burns out.

DIRECT AND INDIRECT SOLAR ENERGY

Solar energy doesn't just mean the sunlight from the sun; there are energies the sun creates indirectly too. For example, during the day land masses are heated faster than the oceans; as a result the air over the land is heated, becomes lighter and rises. The cooler air above the oceans then rushes in to replace the rising air above the land areas, causing air currents, or wind. At night the process is reversed; the land areas cool faster than the oceans, with the air above the land moving towards the oceans, creating wind in the opposite direction. A few hundred years ago it was realised that we could make use of these air currents or wind by using windmills, to pump water, grind wheat and do other chores. Today's successors of the windmill are called wind machines, but they are based on the same principle. Most wind machines are used to create electric power, though some are still used for pumping water and grinding grain, especially in Third World countries.

Another example of indirect solar energy is hydro-electric power. Solar radiation causes the water from oceans, lakes and rivers to evaporate, forming clouds. Winds carry the clouds towards the land areas. As they move over the land masses and meet mountains, the clouds rise higher and higher; then, as they cool, the water vapour condenses and comes down in the form of rain, snow, sleet or dew. Some of this precipitation eventually forms streams and rivers, which empty into lakes and oceans, while some of the water will seep into the ground and flow downhill as groundwater.

In the past, whenever fast-moving streams or rivers were discovered, man diverted them through watermills in order to harness the energy. Nowadays, huge dams are built to collect vast amounts of water. This is then channelled through giant turbines, which produce electricity.

There are some other forms of indirect solar energy that aren't as familiar as the ones we've just mentioned – ocean thermal energy, ocean currents, and waves, the first two being produced by the temperature differences in the oceans, while the waves are produced by winds. Tides are also considered to be indirect forms of solar energy, since they are caused by the gravitational pull of the moon, a part of the solar system.

Indirect forms of solar energy are environmentally sound, although we do have to be careful where we locate dams so that we do not endanger various plant and animal species, or human life.

THE SHORTCOMINGS OF SOLAR ENERGY

Although solar energy in its direct and indirect forms is environmentally compatible, it also has shortcomings. It is not as convenient to use as petroleum or natural gas. For example, we cannot use wind energy to fly large aircraft. And although it is true that there are some small lightweight cars that run on the electricity from solar cells attached to them, they are only usable in wide open spaces, like the countryside, and only in the daytime on sunny days – we cannot pump solar energy into a car.

In the temperate zones the sun shines an average of six to seven hours out of every 24. In the tropical regions the availability of the sun increases a little – seven to eight hours a day. As we move further north and south from the temperate zones, days become longer during the summer and shorter during the winter, but the intensity of the sun is less because of the greater distance the sunlight has to travel through the atmosphere; for example, at the extreme polar regions days are six months long, but the sunlight is weaker than in the midwestern United States, or central Europe, and weaker still than it is at the Equator.

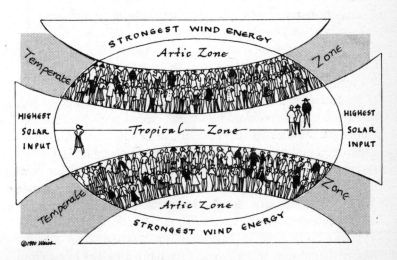

Direct and indirect forms of solar energy are, in general, not available near the consumption centres.

Solar energy has yet another shortcoming, one that we have examined already. Its direct and indirect forms are, in general, not available near the centres of consumption. Solar energy is strongest in the tropical and subtropical regions, while the major consumer groups of energy are in the temperate zones. Wind energy is the strongest in the arctic and subarctic regions, and less available in temperate zones. It is the same with hydropower, ocean-thermal, wave, current and tidal energies; in general, they are most available far away from the areas where they are most needed.

THE NEED FOR A GO-BETWEEN

In short, there are certain times and places when direct and indirect forms of solar energy are not available, and even when they are they may not be in a form that factories, homes and transport systems can use. Somehow therefore, we have to store the energy of the sun, wind, waves, ocean, heat, tides and currents as and when they are available. In other words, we need an energy store that acts as a go-between (or intermediary) between solar energy (direct and indirect) and the consumer.

This go-between must satisfy the following conditions:

- It must be storable.
- It must be transportable.
- It must be a fuel for use in transport systems, homes and industry.
- It must be clean.
- It must be inexhaustible.

To meet all of these requirements, hydrogen is the best. Hydrogen does not produce any of the chemicals that cause the greenhouse effect, nor does it produce any of the chemicals that cause smog or acid rain, except nitrous oxides, and these can be controlled. Indeed, if a device called an electrochemical fuel cell is used, the nitrous oxides are eliminated completely. All it does produce are eletricity and water vapour. Hydrogen is thus the cleanest fuel.

Hydrogen is also an efficient fuel; it can be converted to other forms of energy (such as mechanical and electrical) more efficiently than other fuels. For example, in cars, hydrogen has an efficiency of 60 per cent, while the efficiency of petrol is only 25 per cent. In other words, hydrogen is two and a half times more efficient than petrol. In supersonic jets, hydrogen is 38 per cent more efficient than jet fuel. This increased efficiency means that less fuel is wasted and you get better fuel mileage.

PROPERTIES OF HYDROGEN

Hydrogen, the lightest, most abundant element in nature, makes up approximately 80 per cent of all matter in the Universe. It is present in all types of fossil fuels, even after they are refined. And it is present in the biggest source of energy of all – the sun. The sun is almost 100 per cent pure hydrogen, its energy resulting from the fusion of hydrogen atoms. The planet Jupiter is almost 100 per cent hydrogen, too – liquid on the surface, then frozen, followed by metallic hydrogen in the core (metallic hydrogen is a solid created under high temperatures and pressures).

On earth, hydrogen is rarely found in its free form; most of it is attached to oxygen to make water. Every two out of three atoms in water is hydrogen, so oceans, lakes and rivers are our hydrogen 'mines'. If we are going to use hydrogen as an energy carrier, we must therefore produce it from hydrogen-rich compounds, in this case water (remembering that when hydrogen is used as a fuel, we get water back).

SOLAR-HYDROGEN ENERGY SYSTEM

Hydrogen complements the shortcomings of solar energy very well, this 'coupling' of solar and hydrogen being called the solar-hydrogen energy system.

In this system, hydrogen is produced by any one of several hydrogen-production methods, using solar energy in its direct or indirect forms, depending on what is most convenient. After it is produced it is sent by pipeline or supertanker to

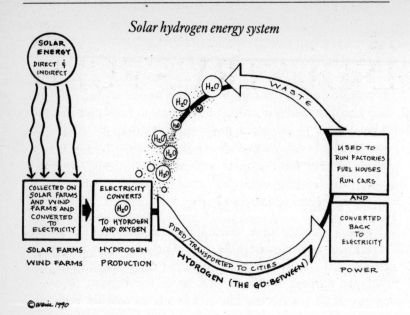

Solar hydrogen energy system

areas where electricity needs to be produced or where it can be used directly as a fuel in heating, cooking or for transportation. In every application where we are using fossil fuels today we could use hydrogen, with many environmental benefits.

Solar energy and hydrogen go hand in hand. They are both environmentally benign; they don't produce pollution, acid rain or the greenhouse effect; and hydrogen is a very efficient fuel, causing less waste.

CONCLUSION

It makes sense to use the forms of energy that are abundant, clean and renewable. The sun and hydrogen have been coupled since the beginning of time and will be producing energy together for a span of time beyond anyone's imagination. For a healthier life it makes sense for us to use solar-hydrogen.

12
INDUSTRY'S FUEL

When solar energy is combined with hydrogen, it can do everything that the fossil fuels can do, and do it better. It can provide energy for transport systems, for heating, cooling, cooking and lighting, not only during the day but also at night and when the skies are cloudy. It could provide the energy needs of people in cities and industrial parks, as well as of those in rural areas. It is a very flexible system that has many applications for the everyday requirements of human society. And it can do all of these things more cleanly and efficiently than fossil fuels do.

INDUSTRY'S FUEL

After the Industrial Revolution, machines started to replace manual labour. This resulted in lower prices for the goods produced, so more people were able to afford more goods. Human society was on the way to affluence. In the two centuries since the Industrial Revolution, factories have become larger and more numerous in order to meet the growing demands of the growing population, and their energy consumption has greatly increased as a result of such growth.

At present, industry runs on fossil fuels but once the infrastructure for hydrogen is ready (tanker trucks and pipelines, for instance), there is no reason why industry could not run on the solar-hydrogen energy system. And it could do this with many environmental benefits – less pollution and greater fuel efficiency, for example.

The first step is always the hardest whenever change needs to be made, but we know that to keep our environment stable we will have to change industry. However, we can make that change as painless as possible. We have several aspects to consider, but first we need to determine how solar energy can become industry's fuel, and then consider some of the applications.

SOLAR-HYDROGEN PRODUCTION

As you read earlier, although solar energy is environmentally benign, it is only intermittently available. In general it is only available on average for about one-third of each day, and even then the intensity of the sunlight varies from a little in the early morning and evening to a lot in the afternoon. That's why we have to store solar energy when it's available at its peak, to use later when it's not.

Using solar energy, hydrogen can be produced cleanly in four different ways – direct heat, thermochemical, electrolytic and photolytic.

Direct heat

In the direct heat method, water is heated to form steam and then steam is superheated to about 1,400°C (2,500°F) or more, at which stage the molecules of steam (superhot H_2O) start splitting apart to form hydrogen gas and oxygen gas. As the temperatures are further increased, the rate at which the steam molecules split apart increases. The same effect can also be achieved by reducing the steam's pressure; in other words, higher temperatures and lower pressures are the best ways to produce hydrogen using the direct heat method.

To make enough hydrogen for industry's use, the temperatures would have to be raised to 2,500–3,000°C (4,500–6,400°F), and to reach such high temperatures the cleanest method of providing heat would be to use the sun. But how to get the heat strong enough? The sun's rays certainly don't hit the earth at that temperature.

The answer is a mirror system that would collect the sunlight and concentrate it into a small area, similar to burning paper using sunlight and a magnifying glass. Large concentrating (parabolic) mirrors would then be used to focus the solar energy on to containers of water. Such a system is called a solar furnace, since it produces very high temperatures – steam without pollution.

There are two problems with the direct heat method, though. One is that the containers used to hold the water will not withstand the high temperatures needed to dissociate the water. There must therefore be an efficient means of cooling them down. The second problem is that, at these very high temperatures, hydrogen and oxygen must be separated, and remain separated, from each other because as they cool down they recombine, forming water again. Research is going on at the French National Solar Energy Laboratories in Odeille, France, to produce hydrogen from water using this direct heat method, and in particular to find efficient and economical ways of separating the two gases.

Thermochemical

But we don't have to have temperatures as high as 2,500–3,000°C (5,000–6,000°F) to split steam. If much cooler steam at 300–1,000°C (600–2,000°F) is passed over powdered iron, the iron soaks up the oxygen, forming iron oxide – rust – and leaving hydrogen. The rust can then be heated to make it release the oxygen, leaving us with unrusted, powdered iron again. By doing this over and over again, with large amounts of powdered iron, we could obtain a supply of hydrogen gas. This is one example of the thermochemical method. Other metals and chemicals are currently being researched to find a cost-effective process.

Electrolytic method

The technology for this method of hydrogen production already exists today. In this method cells similar to the cells in the battery of your car are used to produce hydrogen and oxygen from water. Each cell consists of two electrodes immersed in an electrolyte of water plus some chemicals that conduct electricity well, and is connected to a direct current (DC) electricity supply. When enough electricity is supplied between the electrodes to cause a current to flow, oxygen is produced at one end (the anode) and hydrogen at the other (the cathode). Instead of splitting steam with heat, we're splitting water with electricity.

Keep in mind that, whether we use the sun to make steam in the direct heat or thermochemical methods, or to make electricity to produce hydrogen in the electrolytic method, we are limited to the times when the sun is available. To use these methods for hydrogen production, research is therefore being conducted to find efficient and clean ways to store the sun's heat so we can produce energy 24 hours a day.

Photolytic method
The last method uses the sun's energy directly to split water into hydrogen and oxygen, without the need for high temperatures or electricity.

Water 'soaks up' minute light particles in the sun's rays, called photons. Then, when the water has absorbed enough photons, it splits into hydrogen and oxygen. This phenomenon is called photolysis.

The photons in the ultraviolet portion of sunlight have the higher energy needed for the direct photolysis of water. However, most of the ultraviolet radiation is absorbed in the upper atmosphere by the ozone layer (even though it is thinning). Consequently not much ultraviolet radiation reaches the earth. This is a good thing in some ways, because if there was too much UV light reaching the earth's surface it would be harmful for all living things. Therefore, in order for photolysis to produce enough hydrogen to use in industry, we need to boost either the sunlight or the splitting of the water. Since making the sun stronger would not be healthy, we have to help the water to split by adding specific metals or minerals that soak up more photons from the sunlight than the water could by itself.

The photolytic method for producing hydrogen is not very efficient, but it is cheaper than the other methods because there are no mechanical parts or machinery - nothing that uses energy.

STORAGE

We've shown you that there are clean ways of making hydrogen, but how can it all be stored for industry to use as it needs? For such large-scale storage, solar-hydrogen would best be stored underground; this is the cheapest option. For example, we could use the spaces left by the depleted petroleum and natural gas reservoirs, once these resources are used up, or the man-made caverns resulting from mining and other underground activities. Such storage systems – on a small scale – are already in use in the UK and France.

DISTRIBUTION

Now we have to get the stored hydrogen to industry. On a small scale, hydrogen can be transported and distributed as a gas in a tanker truck, but for industries requiring lots of energy, pipelines are the most economical way to transport and distribute large quantities of hydrogen.

In fact there are already hydrogen pipelines operating in parts of the United States and Europe, and as a result of these pipelines a wealth of experience exists in the safe operation of hydrogen distribution. This experience is important, because if energy needs to be sent a distance of over 2,000 miles it is cheaper to send it through a pipeline than to convert it to electricity and send it by power lines. Electricity, once it is sent through power lines, must be used or it is wasted; hydrogen, on the other hand, remains in the pipelines until it is needed; a considerable advantage.

Another advantage of hydrogen is that it doesn't need land for power lines. For instance, think of an industrial complex, receiving its electrical energy via a high-voltage power line. The string of power lines from the transmission station to the industrial complex would use about a 100-metre (300-foot) or wider right-of-way, extending miles and miles, and there would also be unsightly towers, pylons and cables causing landscape pollution. However, the same industrial complex could be supplied with hydrogen to meet all its energy needs: electrical, by the conversion of hydrogen to electricity through a fuel cell at the complex itself; mechanical, using hydrogen gas in their delivery vehicles; and heat, heating the buildings with hydrogen gas heaters. And all this hydrogen could be supplied via a single underground pipeline.

WHAT HAPPENS TO THE OXYGEN?

In the hydrogen energy system in addition to hydrogen, oxygen will be produced from water as a by-product. The oxygen produced could be kept in containers and/or trans-ported by pipelines to industries which use oxygen. Or it could be released to the atmosphere. When hydrogen is burned it needs oxygen which is taken either from the stored oxygen or from the atmosphere. Hence, just as hydrogen is completely recycled and earth's hydrogen content will not increase or decrease in the hydrogen energy system, the oxygen is also completely recycled and the amount will not increase or decrease.

For some applications (for example, surface transportation or air transportation), the oxygen needed will be obtained from the atmosphere, and hence the corresponding amount of oxygen – when hydrogen and oxygen are produced from water – should be released to the atmosphere. In other applications (for example, in rockets and spacecraft) oxygen and hydrogen are carried in liquid form in separate containers, since there is no oxygen in space. Also, it is expected that for large fuel cell power plants, pure oxygen would be supplied by pipeline. This will increase the efficiency of the fuel cell power plant compared with when oxygen is obtained from the atmosphere.

Of course, oxygen molecules under sunlight, especially ultraviolet light, can first breakdown into atoms and then combine with other elements and produce new molecules, or with other oxygen molecules to produce ozone. But this process is going on all the time within the 21 per cent oxygen content in the atmosphere, and is relatively small. In the hydrogen energy system, this percentage will not change much. There may be very small increases and decreases in the proportion of oxygen in the atmosphere, but the average would remain about 21 per cent.

INDUSTRIAL UTILISATION

Once we have produced solar hydrogen and solar (direct and indirect) electricity, or solar-hydrogen electricity, then all the energy needs of industry could be met. Industries using fossil-fuel or nuclear-generated electricity, and industries using natural gas or fuel oil today, will be able to use solar electricity and solar hydrogen to keep their plants running without polluting the environment.

THE CHANGE TO SOLAR-HYDROGEN ENERGY SYSTEM
...WOULD THAT FIRST STEP *REALLY* BE SO HARD?

Small industrial complex receiving electricity by high-voltage power lines. (Notice the wide swathe of land around the power lines.)

Same small industrial complex supplied with hydrogen to meet its energy needs. (Hydrogen supplied by underground pipeline.)

13
ENERGY FOR HOMES

SOLAR-PRODUCED ELECTRIC POWER

We saw in the last two chapters that direct and indirect solar energy can be used instead of fossil fuels to meet all our energy needs, including the production of electricity. Some of this electricity could be used to split water into hydrogen and oxygen with a machine called an electrolyser, and the hydrogen could then be used as a fuel and for the production of electricity when sunlight (or the indirect forms of solar energy) is not available – at night, for instance.

Hydrogen can be converted to electricity using three different systems – gas turbines, steam turbines and fuel cells. Today we use gas turbines for mechanical energy and to produce electrical energy (when the turbine is coupled with a generator). In either case, natural gas is used, emitting carbon dioxide and other pollutants into the atmosphere; but, were the gas turbines to be run on hydrogen gas instead of natural gas, they could produce the same amount of energy cleanly and efficiently.

We get a lot of mechanical and electrical energy from steam turbines. But the steam turbines used today for the production of electricity operate on coal or fuel oil which emit pollutants that add to the greenhouse effect, or nuclear fuel, which presents its own problems (see Chapter 8). Steam could be produced much more cleanly by burning hydrogen in pure oxygen. However, it is extremely hot – about 3,000°C (5,500°F), or 10 to 20 times hotter than the highest setting on a conventional oven. The faint blue flame we see when the space shuttle takes off is that same high-temperature steam produced by burning hydrogen and oxygen. Of course, no

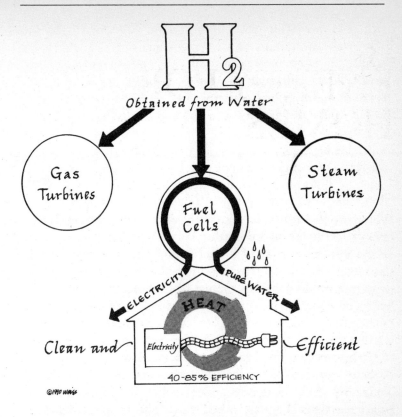

materials could stand such temperatures for long unless they were cooled down, so the rocket engines have to be cooled by circulating liquid hydrogen at −253°C (−423°F) around the engine nozzle. Similarly, in power plants producing electricity, water is added to the high temperature steam to bring the temperature down to a manageable level.

There is also a third method of converting hydrogen into electricity that is unique to hydrogen. Fuel cells use hydrogen as a fuel, combining it with oxygen from the air to produce electricity. The only by-product of this process is water, and it is water, of course, that is the source of hydrogen in the first place, so the process is sustainable and clean.

The electricity the fuel cell produces can be used for domestic and industrial purposes, although a converter is needed to use the electricity in our present power systems. Fuel cells come in many different sizes. A small one, about the size of a house air conditioning plant, would meet all the electrical requirments of a house or apartment. Larger ones could supply factories, shops etc. Depending on the type of fuel cell, its efficiency can range anywhere from 40-85 per cent. In addition to electricity, fuel cells also produce heat, which can be put to use – water heating, space heating and/or drying.

In Japan a 4.5 megawatt (MW) fuel cell, big enough to power a town of 10,000 people has been operating successfully since 1984, and in the United States several 40 kilowatt (kW) fuel cells have been tested in houses and apartments, with efficiencies of up to 70 per cent. Further research work may increase efficiencies still further.

Another advantage of fuel cells is that there would be no need for electricity transmission systems, as every house, building or plant could produce its own electricity through a fuel cell. This would also do away with unsightly electricity pylons and lines, a source of landscape pollution.

Hydrogen is the best fuel for fuel cells. If this hydrogen is obtained from a fossil fuel or an alcohol fuel like methanol, then we will still have the pollutants that cause the greenhouse effect, the acid rains and the smog. But when hydrogen is obtained from water, we don't get those pollutants. One day, every house, factory, shop, office block, etc. could be powered by its own fuel cell, but we must ensure that the hydrogen used to fuel the cell is obtained from the safest source – water.

PRACTICAL APPLICATIONS

Space heating and cooling

Homes, offices and plants need heating and cooling to create the right temperatures for us to live and work in comfortably. This could all be achieved very easily with the solar-hydrogen system. Once the electricity has been generated, it could then be used for heating through electrical resistance heaters or heat pumps, for cooling through conventional air conditioning systems, and for heating and cooling through reverse-cycle air conditioners.

Wall units for space heating have already been designed using a system called catalytic combustion. This system utilises a unique property of hydrogen; that of flameless burning. Air enters the system through the surface of a catalytic plate made of a porous ceramic with traces of platinum. Hydrogen is taken in through the back of the plate and flameless combustion takes place inside producing heat for the room. The plates face into the room and can be hung on or built into the walls, and they can be concealed by panelling materials for an acceptable appearance. Such room heaters would operate at relatively low temperatures and are extremely efficient – close to 100 per cent.

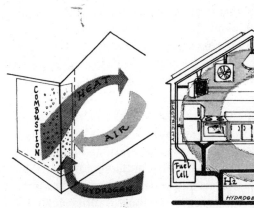

CATALYTIC COMBUSTOR
Close to 100% efficiency

HYDROGEN HOME

Another way of using hydrogen for heating and cooling is a larger version of what the astronauts use in their space suits to stay warm and/or cool – a hydride heat pump. Hydrides are metals or alloys that interact with hydrogen, as sponges do with water. These hydriding materials, or metal sponges, soak up hydrogen and are temperature sensitive, releasing, or 'wringing out', the hydrogen when they get warm and soaking it back up as they cool down. These properties are being used to build heating and cooling systems in the United States and Japan, and we may see them marketed in the near future. A significant advantage of using this system is that it doesn't use

93

CFCs as present heating/cooling systems do, and therefore is not a threat to the ozone layer.

Cooking

Another one of our basic needs is to cook food. We have come a long way since the days of our ancestors, and cannot be expected to give up our ranges, stoves, ovens and microwaves, so a clean, efficient form of energy has to be available to provide this standard of living to which we have grown accustomed. The solar (direct or indirect) hydrogen system can provide it.

The obvious method is to use hydrogen instead of natural gas in conventional gas appliances. Since the hydrogen flame is almost invisible, it would be a good idea to add a small amount of hydrocarbons or other impurities to give the flame a colour – a means of ensuring safety.

The second option is to use conventional electric cooking appliances with the electricity produced from solar-hydrogen through a fuel cell.

But there is a third method – to use the same catalytic combustors that can be used in wall heaters. The primary advantage of using catalytic combustors is that heat can be provided without a flame and at lower temperatures. The catalytic combustors are safer, and transfer more heat to pots, pans and food. Catalytic hydrogen combustor efficiencies for cooking are 85 per cent, while hydrogen used in conventional gas burners is 70 per cent efficient. Both of these methods are better than natural gas burners, which are only 60 per cent efficient.

In tropical and subtropical regions direct solar radiation could be used for open-air barbecues and cooking; there are solar cookers available now which concentrate the solar heat in areas where cooking pots or food are placed.

Lighting

Illumination is another important basic need, not only for our houses and other buildings, but for street lighting too.

During the daytime, the necessary light is usually provided by the sun especially in open spaces and in single storey homes with windows on all sides. Sunlight is the most natural light for human activities, the sun's energy-giving radiation having nurtured the life on this planet in the first place.

But when it comes to high-rise apartments and office buildings, many of the units have little or no daylight. Japan has taken the lead in providing natural sunlight for such buildings; they install 'sunlight pipes'. Starting from a point where sunlight is available, such as the roof or a sun-facing side of the building, and through the use of hollow pipes coated with a reflective surface and with sharp bends containing mirrors, they direct sunlight into interior rooms and corridors. Sunlight carried through this unique system can even be strong enough for people to sunbathe in their bedrooms. However, unless Japan's idea catches on, solar- and/or hydrogen-produced electricity is the most convenient energy carrier to power systems using electric light bulbs.

Water heating

In homes, offices, public places and factories, we need hot water. In the tropical, subtropical and temperate regions of the earth, solar radiation can be used directly for producing hot water through solar water-heating systems – solar water-heaters have been the first economical application of solar energy. They are quite popular in Mediterranean countries, in the southern United States, and in the sun-belt countries.

But for those regions that don't get so much sunlight, a conventional hot water heater, using hydrogen instead of natural gas, would be a convenient method of obtaining hot water. The efficiency of a natural gas hot-water heater is 60 per cent, while that of the hydrogen hot-water heater is 70 per cent (the higher the efficiency the more heat and less waste), and there are other methods being developed using hydrogen that could heat water at an efficiency of nearly 100 per cent such as catalytic combustors.

Water supply

Of course, one of the basic needs of human society is water. In rural areas, if piped water is unavailable, then underground wells are used. Solar hydrogen energy could be used in several ways to pump up the underground water, one being to use solar and/or hydrogen electricity with a conventional water pump, while another is to couple a conventional pump to a hydrogen-powered internal combustion engine when electricity isn't available. A third option, if strong winds are available, is a wind-operated pump, and a fourth, which is being developed at the moment, utilises hydrogen-hydride systems using solar and/or waste heat.

Refrigeration
In the old days we kept our food in screened pantries that provided an airflow over the food and thus a degree of cooling. Then we used ice boxes; ice was bought from the ice man, who sold ice manufactured in a central plant or ice kept underground from the previous winter. Nowadays we have refrigerators to keep our groceries fresh.

If solar-and/or hydrogen-produced electricity is available, then one way of providing refrigeration is to use conventional electric refrigerators. A second method of refrigeration would be to use hydrogen instead of natural gas to heat the refrigerant in conventional natural gas-powered refrigerators; these units operate without moving parts, thereby resulting in quiet operation, long life and low maintenance. A third method of refrigeration would be to use the catalytic combustors; this method would heat the refrigerant without a flame, further improving energy efficiency and conservation. A fourth alternative would be to use the hydrogen-hydride system, which would eliminate the need for CFCs and help to protect the ozone layer.

Home appliances and equipment
Many household appliances, equipment and gadgets – sewing machines, dishwashers, electric toothbrushes, computers, saws, drills, radios, televisions, sound systems, etc. – are dependent on electric motors or electronics to function. This equipment could all be run on solar- and/or hydrogen-produced electricity, with none of the pollutant by-products caused by fossil fuel-produced electricity.

But some of these appliances and gadgets – clothes dryers,

stoves, refrigerators and hot-water heaters – might also run on a fuel. In such cases heat energy, produced either through a flame or the catalytic combustion of hydrogen, could be used.

THE SOLAR-HYDROGEN HOME AND CAR

In the future, when the nations of the world are fully converted to the solar-hydrogen energy system, we envisage that there will no longer be landscape 'pollution' such as electric power lines, criss-crossing the countryside and city streets. Energy for all human needs will come through solar (direct or indirect) systems producing hydrogen. In urban areas solar hydrogen power will come from the so-called solar farms and/or plants operating on direct and/or indirect forms of solar energy, depending on the location and time.

Are you ready for this ... according to the National Wildlife Federation (based on statistics using electricity generated by coal) – your personal contribution to global warming is ... *12.80 lbs of carbon dioxide/hour from the toaster oven – 12.80 lbs of CO_2/day from the refrigerator ... not to mention the other appliances in the kitchen and the rest of the house!*

Hydrogen supplied by central hydrogen-producing plants will enter the home or the building through underground pipelines, and will be used wherever we use natural gas or fuel oil today (for example, for heating, cooking, drying, etc.). Various kinds of fuel cells will be used to meet the electricity needs. The family car could be filled with the hydrogen gas from the household pipeline for in-town driving (of course, fuel stations would still be needed for long-distance driving). Experiments show that cars fuelled by hydrogen need very little servicing, since hydrogen does not contain any corrosive chemicals, nor any carbon to build up in the combustion chamber. As a result, visits to the garage would be less frequent than we are used to today. These additional savings would be yet another advantage of the proposed solar-hydrogen energy system as compared to the present fossil fuel system.

Hydrogen is thus a versatile way to use direct and indirect solar energy cleanly and efficiently at night (or anytime), without waking up to smelly, hazardous pollutants in the morning.

What you sleep with is your choice.

14
PLANES, TRAINS, CARS AND SHIPS

Yes, planes, trains, cars and ships can all run on solar power, or on solar-produced hydrogen. Hydrogen's many unique properties – being the lightest and the fastest burning fuel known – make it an excellent fuel for rockets and engines, indeed it is ideal for many forms of transportation.

SPACE TRAVEL

Space ships must travel long distances to explore the heavens beyond our planet. In addition they must move at very high speeds, something like 30 times the speed of a subsonic jet (or 25 times the speed of sound) in order to escape the earth's gravitational pull to orbit the earth or travel to the moon and beyond. To achieve the desired speeds they must have powerful engines; rocket engines or jet engines. These engines consume large amounts of fuel, so space ships must have the capacity to carry all the fuel, as well as the oxidiser, oxygen, with them, so that they can accomplish their missions. (There is no oxygen in space, so space ships must carry the oxygen necessary for their trips.) Of course, in order to save weight and therefore fuel, they must carry the lightest possible fuel. Hydrogen is ideal because it is the lightest known element in the universe, and it can be compressed or liquefied to save space. And because it is so light, it burns quickly making it the fuel of choice for space programmes. The National Aeronautics and Space Administration (NASA) of the United States is the biggest consumer of hydrogen in the world; it is the principal fuel for their space missions. The Europeans, Soviets and Japanese also run their space programmes on hydrogen fuel.

Without hydrogen NASA could not have accomplished the amazing feat of putting a man on the moon and bringing him back safely. It should also be mentioned here that hydrogen has not been implicated as a cause of the Challenger disaster in 1985. The accident was caused by flames leaking out of the solid fuel boosters, the seams of which were defective, that then penetrated into the hydrogen and oxygen storage tanks producing an explosion.

NASA is now planning an ambitious space station. It will be carried piece by piece into space by the space shuttle sometime in the late 1990's. The main fuel for these trips will be hydrogen. In addition, hydrogen will be used to power the space station through the use of hydrogen-oxygen fuel cells. You will remember from the last chapter that hydrogen fuel cells, in addition to providing electricity to power the various operations of the space station, will also produce pure water.

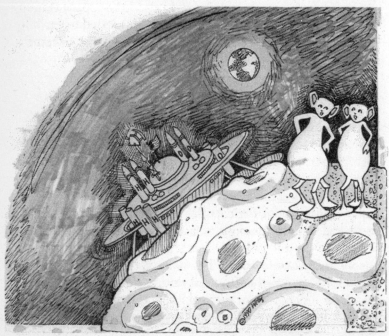

I can't figure out why they use this hydrogen/oxygen fuel system exclusively here and they have such a hard time adapting to it on their own planet.

Electricity produced by solar cells attached to the space station will be used to split water into hydrogen and oxygen, which will in turn be used to power the fuel cells during the time when the space station is in the earth's shadow. As the solar panels would not be able to produce electricity when the station is not in the sun's rays, hydrogen is an effective way of storing fuel for electricity generation at any time.

NASA and the Soviets are also planning a joint mission to Mars. The spacecraft must carry all the supplies needed for a trip which may last as long as eight months to a year and, of course, it must be as light as possible to escape the earth's gravitational pull, attain orbit and travel to Mars. This spaceship will also run on hydrogen. It will be assembled and prepared for its long trip at the space station, and the components of the space ship – its fuel, its oxidiser, its provisions and its crew – will be transported to the space station by the space shuttle. It will return to the space station, and the crew and Martian treasures will be carried to earth by (you guessed it!) the shuttle. Such trips could not become a reality without hydrogen.

AEROPLANES, AEROSPACEPLANES AND BLIMPS

Because of its light weight and excellent combustion characteristics, hydrogen is the ideal fuel for aeroplanes and aerospaceplanes, which will fly both in the atmosphere and in space. On April 15, 1988 the first passenger plane flew on a hydrogen-fuelled engine near Moscow. The Tupolev 155 (equivalent to an American Boeing 727) was equipped with two engines – one running on hydrogen, the other on jet fuel (a fossil fuel) – a liquid hydrogen storage tank, and a hydrogen supply and control system. The plane took off and landed on jet fuel, but hydrogen was used during the cruising phase of the flight. The various aeronautical establishments of the Soviet Union and the Tupolev Institute are now working on the design and development of an all-hydrogen supersonic passenger plane, which will be called the Tupolev 204.

On June 17, 1988, two months after the flight of the Soviet jet, Bill Conrad, a retired Pan American pilot, flew a hydrogen-fuelled single-engine plane in Fort Lauderdale, Florida. The flight lasted only 36 seconds, but the fact that it was fuelled entirely by hydrogen in take-off, flight and landing established a new record.

Actually, Mr Conrad's plan was to taxi down the runway and return to the starting point, then take off, fly a few times around the airport and land, all on hydrogen fuel. Because hydrogen fuel is more efficient than normal fuel, the plane suddenly lifted off the ground while taxiing. Mr Conrad immediately reduced power, put the plane back on the runway, continued in his taxiing mode and returned to the starting point, ready for the flight. The officials from the Civil Aeronautics Board and other recording agencies told Mr Conrad that he had already established a record and there was no need for him to fly again.

Because of the importance of space as the next frontier, in 1987, President Ronald Reagan charged NASA and the US Air Force with the development of the National Aerospace Plane, which will be capable of speeds of five to 25 mach (five to 25

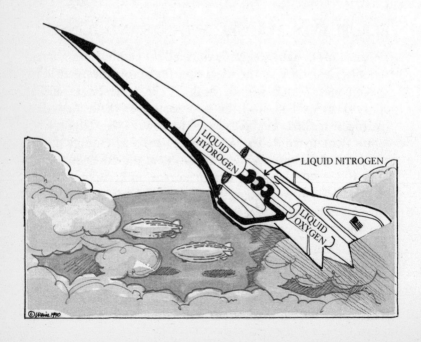

times the speed of sound). The passenger version of this plane is expected to fly from New York to Tokyo, Japan, or from Los Angeles to Sydney, Australia, in two hours, and has been nicknamed the 'Orient Express'.

The space version of this plane will be able to take off like an aeroplane, escape the gravitational pull of the earth, travel outside the earth's atmosphere in space, then descend safely through the atmosphere and land on a runway just like an ordinary aeroplane. Its engines will use oxygen from the air to keep the fuel burning during its flight within the atmosphere and will use stored liquid oxygen in space. The fuel selected for the National Aerospace Plane is, of course, hydrogen. The first flight is scheduled for 1996.

Other countries are also developing versions of the aerospace plane. The British version is called HOTOL. Sänger, the German version will consist of two planes, one piggybacked on the other. The larger (mother) plane will be capable of speeds of up to five mach. The smaller plane, running on liquid hydrogen, will then take off from the back of the mother plane, fly into space, then return through the earth's atmosphere and land on a conventional runway. The mother plane, of course, will land after the aerospace plane has taken off.

The Japanese government has decided to develop a hypersonic aerospace plane capable of speeds of five to seven mach. The plane will take about seven years to develop, and they are planning its first flight before the end of this century.

Aerospace companies and manufacturers of jet engines also have plans for air and space travel in the 21st century. It is expected that subsonic, supersonic and hypersonic passenger jets will be flying on hydrogen early in the next century. Studies by the Lockheed Corporation indicate that subsonic jets running on hydrogen would be more efficient than those running on jet fuel. Supersonic jets such as the present day Anglo-French Concorde, would also be more efficient if they ran on hydrogen than if they continued to run on jet fuel.

In addition to being the ideal fuel for planes and space ships, hydrogen, as the lightest element, is the ideal gas for blimps. The engines of these airships could use some of the hydrogen, which is stored in the balloon for uplift, as fuel during flight. There are plans by some aviation companies to build blimps to carry large cargos cheaply across continents and oceans. Since we now know how to handle hydrogen safely, it would be possible to design and build blimps which would not cause accidents, such as that of the famous Hindenburg disaster.

TRAINS

Although many nations, such as Britain and France, operate electrified or partially electrified rail systems, in the United States and many other countries the railways are mostly dependent on diesel fuel at the present time. Because we will be running out of oil in the future, and because of the environmental problems caused by oil products, other fuels, among them hydrogen, are presently being examined as feasible alternatives.

Railways consume about 1.5 per cent of the fossil fuels used in the world, mainly in the form of diesel fuel. However, they carry about 50 per cent of the bulk goods that are moved across land, reflecting high energy efficiencies as compared to other methods of transport.

Because diesel fuels will become a scarce commodity, and in order to protect the environment, the railways are examining various future options. All electric railways (of course, to run on solar-produced electricity in the future) are a solar-hydrogen energy system candidate, and they are already widely used throughout the world. However, electrification of railways in the United States must be seen in terms of the cost to make the change. A changeover can only be justified where there is heavy railway traffic; for example from Washington, D.C., to New York and Boston.

Whenever and wherever the large investment for electrification cannot be justified financially, hydrogen-fuelled locomotives will be a practicable alternative. Studies by such groups as the Federal Railroad Administration, the American Association of Railroads, locomotive manufacturers and railway companies indicate no major technical barriers to a conversion to the use of hydrogen as a fuel.

In fact hydrogen has been on the railway scene for a number of years. It is routinely moved across the United States as a super cold or cryogenic liquid in specially designed insulated rail tank wagons. This clearly demonstrates the feasibility of handling liquid hydrogen safely under field conditions.

Canadian railways are also working together with locomotive manufacturers towards a demonstration project of a hydrogen-diesel-engine/electric-powered train to run between Toronto and Winnipeg on the Canadian-Pacific railroad.

Hydrogen fuel-cell power plants would also be very practicable for locomotives. Fuel cells may not be the power plants of choice in the near future for automobiles, because at the moment they are bulkier than internal combustion engines. But, since size and space limitations are not as constraining on trains, hydrogen fuel cells would make an excellent alternative to hydrogen-fuelled diesel engines. In fact, they would be a more attractive alternative to replace the diesel locomotive, because they are cleaner and more efficient.

CARS

In the past we had donkeys and horses for personal transportation, but today, particularly in the richer Western nations, many people use the automobile. We use cars to commute to our jobs, for shopping and for leisure: we use cars for everything! Unless we convert our cars to solar hydrogen, we could not have a completely integrated solar-hydrogen system.

Hydrogen has a unique property that makes it ideal for automobile transportation: it burns lean. Petrol and other petroleum-based fuels, natural gas and other hydrocarbons, burn rich. This means that there must be a rich mixture of fuel with air for hydrocarbon fuels to achieve good combustion. For example, when you stop your car at traffic lights or when you slow down, the flow of petrol into the engine is not reduced with the reduced demand in power. Therefore, a lot of fuel is wasted. If the flow of petrol into the engine were reduced, then combustion would cease and the engine would stop running.

However, in the case of hydrogen, when the car stops at traffic lights or slows down, the hydrogen flow into the engine can be reduced but the combustion will continue because hydrogen burns so lean. Thus hydrogen is not wasted when it is not needed. Because of this characteristic, hydrogen-fuelled automobile engines are 60 per cent more efficient than their petrol burning equivalents, as well as being much cleaner.

When less bulky fuel-cell systems are developed, then cars could run more efficiently than they do with an internal combustion engine, either hydrogen powered or petrol powered. Hydrogen fuel-cell-powered cars would be two to three times more efficient than petrol-burning cars.

Around the world there are already experimental cars and buses running on hydrogen-powered internal combustion engines. University of Miami scientists, Adt and Swain, converted one of the first cars to run on hydrogen in 1972. A hydrogen-fuelled city bus started running in Riverside, California, in 1977 and was used to study the feasibility of using hydrogen to relieve the smog problem in the Log Angeles basin. A fleet of cars developed by Daimler-Benz (also known as Mercedes-Benz), has been successfully operating on hydrogen in Berlin since 1985.

The only obstacle to the widespread use of hydrogen-fuelled cars is the price of hydrogen today. It appears to be two to three times that of petrol. However, looks can be deceiving; when you consider the environmental damage caused by fossil fuels and the higher efficiency of hydrogen, then hydrogen becomes quite feasible.

As mentioned earlier, hydrogen can be stored in three different ways: as a pressurised gas in high-pressure containers, as a liquid in insulated containers, and in chemical combination with some metals and alloys called hydrides. This last method of storage is unique to hydrogen. No other fuel can be stored in such a way. Hydriding metals and alloys absorb hydrogen as a sponge would absorb water. In other words, hydriding metals or alloys can store hydrogen very compactly.

Research and development work is being continued around the world to determine which storage method would be best for hydrogen-fuelled cars. There are demonstration projects using each of the mentioned methods. For example, the Miami car used pressurised hydrogen gas, and the Riverside bus used hydrogen stored in hydriding materials. Some passenger demonstration cars in Germany use liquid hydrogen storage.

SHIPS

Hydrogen will make an ideal fuel for ships as well, especially because of its hydride storage properties. Ships need ballast to lower their centre of gravity and thus remain stable. Ballast is usually an additional weight, such as some heavy metal. In the case of hydrogen-fuelled ships, hydriding materials could be used to kill two birds with one stone – to provide both ballast and to act as a fuel tank. Of course, if needed, pressurised gas storage and/or liquid storage could be used as well. The stored hydrogen would then be used in gas turbines, internal combustion engines or (most efficiently) in fuel cells to propel the ship and provide power for other needs.

It is proposed that supertankers could be used to transport hydrogen from solar belt countries to industrial countries in the form of liquid hydrogen stored in spherical tanks similar to today's liquid natural gas (LNG) or liquid petroleum gas (LPG) tankers. Hydrogen stored in hydriding material placed along the keel of the tanker would still be necessary to provide ballast since hydrogen itself doesn't weigh very much (about a tenth of the weight of oil) and is not heavy enough to keep the ship stable.

Hydrogen also makes an ideal fuel for submarines. Nowadays, non-nuclear submarines run on diesel engines when travelling on the ocean surface – which produce atmospheric pollution – or on lead-acid electric batteries under water. The submarines being developed in Germany today run on hydrogen-oxygen fuel cells. Hydrogen is stored in hydrides which also provide the necessary ballast, and oxygen is stored in liquefied form in insulated containers. The

submarine runs on hydrogen and air when travelling on the surface, and on hydrogen and liquid oxygen under water. It is much quieter than diesel-fuelled submarines and has a longer range under water – both important military considerations. It does not have dangerous chemicals such as those in lead-acid batteries, and does not produce pollutants, so is healthier for navy personnel and the environment.

The German navy is very impressed with the trial runs of the first hydrogen submarine. They have reported that hydrogen fuel will be used in their next generation of submarines.

There are also ideas to build large hydrogen-fuelled 'merchant submarines' for commercial cargo carrying purposes. They would be quite efficient and would be able to avoid stormy seas by diving beneath the wave zones.

Of course, the biggest advantage of hydrogen-fuelled ships, hydrogen tanker ships and, of course, the solar-hydrogen system will be that if an accident occurs, there won't be anything like the oil spills which have so damaged the environment, because hydrogen evaporates quickly and disappears. It doesn't stay on the water like oil. If there is a fire, it will only produce water vapour, not poisonous carbon monoxide or suffocating toxic smoke that would cause unconsciousness in a fire victim or a fire fighter.

In April 1989, the tanker *Exxon Valdez* ran aground on a reef in Prince William Sound, Alaska, spilling more than 10 million gallons of crude oil. It spread over more than 900 square miles of water, having a disastrous effect on the environment and killing millions of fish, mammals, birds and plants living in the area. Scientists who have studied previous spills expect extensive damage, with some effects lasting 10 to 20 years.

We cannot guarantee that such disasters will not happen again. But when we convert to the solar-hydrogen energy system, we will not have to worry about accidents like the *Exxon Valdez*.

15
WHEN?

In this book we are advocating what is probably the biggest change in our technological history – changing our fuel. However, breaking away from polluting fossil fuels will probably not be carried out on the initiative of governments, but because of pressure from the voter – you. And there are a number of hurdles to be overcome before this pressure can bear any fruit:

- Teaching people about solar-hydrogen power.
- Changing the spending priorities of governments.
- Solving various research problems.
- Developing an emergency spirit towards tackling environmental problems.

GETTING OVER THE HURDLES

Whenever people contemplate what they see as beneficial change, they invariably want the change to happen as quickly as possible – instant gratification. Many people expect that, when they vote, whatever they have voted for will happen almost at once, or at the very least by the following year. What we have to tell you in this chapter is that changing our fuel system in such a short length of time is not possible, although of course a beginning can be made.

What we are saying is that new technologies take time to be implemented. Even though some products that use solar energy are available today (solar calculators, patio lights and rooftop panels for hot water), we need more research in this area for the move up to the large-scale production of hydrogen on a worldwide basis. Even though hydrogen has been used as a fuel in NASA's space programme for many years, much more research needs to be carried out to bring this technology to a common market, to make it more affordable and usable.

For the space programme the goal was not to produce rockets at the cheapest price, but to produce rockets that could orbit the earth and eventually take a man to the moon and back, whatever the cost. The second generation spacecraft, the space shuttle, was designed with cost-consciousness in mind; it had to be able to be used many times and so had to be able to withstand the extreme temperatures and force experienced when leaving and re-entering the earth's gravitational pull. Naturally, the general population doesn't require vehicles as complex as a rocket or the space shuttle; instead it requires an affordable mass-produced vehicle that is reliable under all kinds of different conditions and that operates pretty much the same in Honolulu as it does in Moscow or Timbuktu.

RESEARCH, RESEARCH, RESEARCH

For hydrogen fuel, research still needs to be conducted into the best ways to store it. For example, will the storage materials survive different weather conditions? How long will they last before they wear out?

Another area to be studied is leakage through the walls of pipelines. When natural gas is transported long distances in pipelines, some of it leaks out along the way. Hydrogen will leak to a greater extent than natural gas because hydrogen molecules are smaller and lighter than natural gas molecules and can slip out through the tiniest cracks. Perhaps the pipes should be lined with plastic coatings, but we don't yet know which ones.

There are many other questions to be answered, but what we do know is that the technology of producing clean hydrogen fuel to replace polluting fossil fuels is here now. Demonstration projects have been run on hydrogen; cars, motorcycles, mopeds, planes and a coal-mining vehicle have all been run on clean hydrogen. But these are all one-time projects; they were built or converted to prove that hydrogen-powered machines work. To mass produce cars, motorcycles, mopeds, planes and all the other machines that use fossil fuels so that, instead, they operate on hydrogen, and to convert the pipeline infrastructure so that hydrogen can be transported, we need to spend hundreds of millions of dollars per year.

Now this may sound like a lot, but it is a trivial sum when we consider the cost of the awesome damage caused by pollution. That cost is estimated in hundreds of billions of dollars per year. Putting right this environmental damage is a task that has to be carried out by governments and/or industry, and who pays for it all? The taxpayer and consumer – us – in higher taxes and prices. There is thus a hard economic logic to persuading governments to shift their spending priorities from fossil energy to new clean energy industries so that we don't pollute the environment and yet still have the energy we need to run our factories and homes.

One of these clean new industries that is rising to the fore is solar power. Solar power in a sense is ready now, though not on a scale available to a mass market; for example, we know it is possible to collect solar energy and convert it to electricity. But many questions remain. For instance, do the solar panels that produce electricity on the rooftop of someone's home work as well when they're producing enough electricity for a city? Do they work in large areas on sunny humid days as well as sunny dry ones? If solar-produced electricity is used to split water to make hydrogen, can that hydrogen then be economically competitive with fossil fuels if produced on a large scale?

THE EMERGENCY SPIRIT

Studies have been made of the time new technologies take to develop, and the results are worrying if you subscribe to the instant gratification view of life. They show that, from the point at which a scientist publishes an idea to the point at which the product is available in the shops takes about 75 years. From the point at which an engineer takes the scientist's idea and builds a model and tests it to see if it works, to the point at which it is available in the shops takes 50 years. But from the point at which a company takes the model and commercialises it to the point at which it is available in the shops only takes about 15 years, maybe less.

However these timescales can change with the circumstances. In wartime, for example, everything gets done many many times faster. In World War II, car factories were changed into tank factories in six months, and weapons were developed and produced at an accelerated rate. An emergency spirit can unify nations. We need that same emergency spirit to develop and produce clean fuels, to convert the factories using dirty fossil fuels to factories using clean solar-hydrogen energy. In short we need an emergency spirit to stop the war we are waging against the environment.

The emergency spirit

16
BUT IS IT SAFE?

The primary energy source under consideration, solar energy, is environmentally the most benign – it is the life giver to this planet earth. But when the safety of the energy carrier, hydrogen, is considered the *Hindenburg* accident is invariably brought into the discussion. People say hydrogen causes explosions and big fires, that it is a very dangerous fuel. However, contrary to this belief, extensive studies have shown that hydrogen is a relatively safe fuel.

For a start, NASA is the world's biggest hydrogen consumer, and has sponsored many studies to compare hydrogen with other fuels, such as petroleum. All of these studies show that, when everything is taken into consideration, hydrogen is the *safest* fuel.

THE *HINDENBURG*

The *Hindenburg* was the biggest passenger airship of its time in 1937. Since hydrogen is the lightest element, its balloon was filled with hydrogen to keep it aloft.

There was another gas available for such purposes – helium. Helium is the second lightest element, and is a noble gas, which means it will not burn. It exists in its natural state in very small quantities, usually mixed with natural gas. Even in 1937 it could have been used to keep an airship such as the *Hindenburg* aloft, but the United States was the only producer of helium at the time. As helium was considered of strategic importance its export was forbidden, the *Hindenburg* had to be filled with the second best choice, hydrogen.

On its eleventh crossing of the Atlantic from Europe to Lakehurst, New Jersey in 1937, it had some 100 passengers and crew members. As people began to disembark, a fire started in the balloon. Nobody knows for sure how the fire started: some say it was because of a malfunction in an electrical contact;

others say it was sabotage (remember this was just before World War II, and the *Hindenburg* was a German airship). In any case, the airship caught fire with the majority of the passengers still aboard.

People panicked – 35 jumped and perished but the others remained in the gondola. Four people died because of fire, but the fire that killed them was in the diesel engines. Incredibly no one was hurt by the fire from the blazing balloon. The reason is that hydrogen flames radiate very little heat; in contrast, fires from fossil fuels (e.g. diesel fuel) radiate very intense heat, due to carbon's ability to get hot and stay hot. If the people on board had come into contact with the flames of the hydrogen fire, they would have been burnt. Fortunately, they were able to make a hasty retreat from the burning airship before the flames could reach them and were saved.

LIGHTER THAN AIR

Besides space travel (where hydrogen is already used as a fuel) the earliest commercial uses for hydrogen will be in planes, particularly in hypersonic aeroplanes that will travel at several times the speed of sound and will be able to cross the oceans in a couple of hours. And sadly, regardless of the precautions taken and the safety features, there will be the occasional unforeseen accidents, on the ground and during the flight.

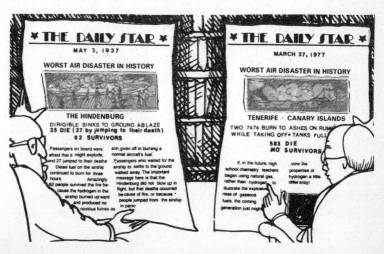

Let us consider a ground accident. In 1977, 40 years after the *Hindenburg* accident, the worst aviation disaster in history took place in Santa Cruz de Tenerife, on the Canary Islands, when a KLM B-747 jumbo jet collided with a Pan Am B-747 on the runway. Both planes burst into flames, their jet fuel burning fiercely, flames and smoke everywhere. Some 583 people perished, mainly from the intense heat and from smoke inhalation. Yet if these planes had been hydrogen fuelled, a disaster of this magnitude probably would not have happened. There would have been no suffocating or breathing difficulties, since hydrogen has no toxic combustion products, and most of the passengers, except those engulfed in flames, would have been saved.

Now let us consider a mishap during the most dangerous time of any flight – after take-off. Today, if there is an emergency immediately after a take-off, the plane either has to circle for hours to empty its fuel tanks, or dumps the fuel, before landing. This is because the landing gear would collapse if a landing was attempted with the full weight of the plane plus a full load of jet fuel (for long flights about 60 per cent of a plane's take-off weight is fuel). But if the plane were fuelled with hydrogen, the weight of the fuel would only constitute 20 per cent of the total weight, a 40 per cent saving. So if there was an emergency right after take-off the pilot could land immediately, avoiding the possibility of further complications. And if any of the hydrogen fuel were dumped, it would evaporate into the air with no poisonous effects; in contrast, any dumped jet fuel pollutes the atmosphere, seas and land.

So hydrogen-fuelled planes would be safer than today's planes in many respects.

IN YOUR CAR

Every year thousands of people die in accidents directly related to petrol or diesel fuel used in road vehicles.

Vehicles running on today's fuels, as we have already seen, produce a chemical cocktail of pollutants. One of these chemicals is carbon monoxide, which, if breathed in, is

Let us consider hydrogen fuel

TEENAGERS FOUND DEAD IN CAR IN MORNING ~ OVERCOME BY CARBON MONOXIDE

THERE ARE 1,800 CARBON MONOXIDE DEATHS EACH YEAR IN THE UNITED STATES ALONE.

TEENAGERS FOUND DRENCHED BY WATER VAPOUR INSIDE CAR IN THE MORNING ~ AN EMBARRASSING SITUATION!

poisonous. Many deaths each year, both accidental and deliberate, occur as a result of people inhaling carbon monoxide fumes from car exhausts; for example, there are 1,800 such deaths each year in the US alone. In contrast, if cars ran on hydrogen, nothing like these tragedies could happen, as the only by-product of the combustion of hydrogen is water vapour. Water vapour, or steam, is not poisonous and does not kill people; it does not affect the nervous system, so people don't become unconscious from breathing it.

If a collision occurs between vehicles running on petrol or diesel fuel, the fuel lines frequently burst, splattering fuel all over the place. In many instances a fire then starts, in which the intensity of the heat, and the poisonous gases, can again prove to be lethal.

As mentioned earlier, the storage system for hydrogen-fuelled cars could be a hydriding metal 'sponge', a liquid hydrogen tank, or a compressed gas storage system. Of the three, the hydriding system is the safest. Even in very violent collisions, puncturing the hydriding tank would not start a big fire; a small flame would merely flicker from the tank as hydrogen was slowly released from the hydriding metal 'sponge'. Some demonstrations show that when bullets are shot at hydriding tanks, only a small flame comes out of the bullet hole; for other fuels, an explosion or fire result.

There are some prototype cars running on liquid hydrogen which have been involved in accidents. One such accident occurred in California, where a car with a full tank of liquid hydrogen was overturned in a collision and the storage tank was punctured. The cold liquid ($-253°C$, $-423°F$) quickly spilled out and evaporated into the air. There was no fire, no explosion.

Cars running on hydrogen stored in pressurised containers pose more danger than those with hydrides or liquid storage; there would be a possibility of fire, but not necessarily an explosion, as explosions only occur when hydrogen and oxygen or hydrogen and air are mixed in a confined space, like a tank. If hydrogen escapes from a broken pipe or a punctured tank, a fire could be ignited by a spark or by the hot surface of the engine, yet people near the fire but away from the flames would not be burned or overcome by toxic fumes. However anyone engulfed by the flames would, of course, be burned.

In conclusion, when all factors are considered, hydrogen is a much safer fuel for cars than petrol and diesel oil.

HOME AND THE KITCHEN

In homes electricity, natural gas, fuel oil or coal are used for heating, cooling and cooking. If for any reason a leak of natural gas or fuel oil vapour occurs in a large concentration anywhere in the house it can be fatal. Furthermore, fires resulting from natural gas or fuel oil produce toxic fumes, as well as high radiative temperatures. Many accidents of this nature can cause injury and death.

But a home running on hydrogen would not be endangered by such mishaps; neither hydrogen itself nor its combustion product, water vapour, is poisonous. Catalytic combustors using hydrogen are the best choice for cooking, since they do not produce a flame and therefore could not cause a fire. If a finger touched a catalytic combustor it would suffer a much milder burn compared to a burn from a natural gas or fuel oil burner, or an electric hotplate; this is because catalytic combustion temperatures are much lower (150°C, 300°F) than the flame temperatures of natural gas or oil (1,400°C, 2,500°F). Rooms heated by catalytic hydrogen burners would be much safer than rooms which use a fireplace or space heater fuelled by natural gas, fuel oil or coal.

Of course, all fuels, including hydrogen, can cause fires. But if a hydrogen fire occurs in a house no one would be burned, unless they were in direct contact with the flames. They would not be overcome by inhaling toxic fumes either, and could quickly and easily escape. They would also be able to help those injured by the flames, since they would not be in immediate danger. However, in the case of fires caused by fuel oil, natural gas or coal, the situation is quite different. The fossil fuel flames radiate intense heat, burning everyone in close proximity, while the poisonous smoke and fumes produced by fossil fuels can quickly suffocate people.

In summary, a hydrogen-fuelled home is safer than a fossil-fuelled home, and even in the event of a fire the residents of a hydrogen-fuelled home stand a greater chance of escaping injury and/or death.

LEAKS

All fuel storage depots and pipelines are subject to leakage, and hydrogen, being the lightest and smallest element, leaks more easily than other fuels, including natural gas. However, hydrogen is a bulkier fuel; for example, it occupies three times more space than natural gas for the same amount of energy. For these, and other reasons, it is not immediately obvious whether hydrogen would be more or less dangerous than natural gas or petroleum under similar leakage conditions.

For example, to compare natural gas and hydrogen from the point of view of potential danger, we could consider the amount of energy that flows through a leak; if the escaping gases were to ignite, the energy content would give a measure of the intensity of the ensuing fire. But since hydrogen contains less energy than natural gas, a higher volume flow of hydrogen would not necessarily mean a higher escape of energy leaking out of the container or the pipeline.

Another factor to take into consideration is that hydrogen flows three times faster from a large crack than natural gas, but from a small crack the flow of hydrogen is only twice as fast as natural gas. This means that if leaking gases from a larger crack were ignited, the damaging forces of the resulting fires would be about the same for both hydrogen and natural gas. For smaller cracks, the damaging force from a natural gas fire would be worse than that of hydrogen. However, when we consider the greater intensity of heat from natural gas fires and the noxious fumes generated, hydrogen is the safer of the two fuels for all kinds of leaks.

Every year millions of gallons of diesel oil, petrol and fuel oil leak from storage tanks and pipelines. They can then enter the groundwater, and make it unsafe for drinking, and if this polluted underground water flows into any lakes and rivers, it will then poison the aquatic life. But this can't happen with hydrogen. If there is a hydrogen leak from a storage tank or a pipeline, it will seep up through the ground, and diffuse into the air.

It is proposed that hydrogen be transported across the oceans in supertankers similar to today's LPG (liquified petroleum gas) and LNG (liquid natural gas) tankers. LPG and LNG are carried on tankers in huge insulated spherical containers, and liquid hydrogen would be carried in the same way. If such a vessel were damaged, or if the tanks leaked, hydrogen would merely evaporate into the atmosphere, which would not harm any living creature or the environment.

This cannot be said for fossil fuels. We all remember the now infamous oil spill of the *Exxon Valdez* that occurred in April 1989 in Prince William Sound in Alaska. It ran on to a reef, spilling in excess of 10 million gallons of oil, with great loss of life to seals, otters and birds; it damaged spawning grounds and nursery areas for salmon and other fish; and it polluted the waters and beaches of Prince William Sound and the Alaskan Gulf. Nothing like this can or could happen with hydrogen because it would have simply evaporated.

So, is hydrogen safe? If we consider all the above facts, we can say that, yes, hydrogen is a safe fuel. But remember, it is a fuel and therefore must be handled with care.

17
WHAT YOU CAN DO

What you can do is spread the knowledge about the new fuel – pure clean hydrogen – and the more complex alternative of converting everything to run on electricity.

There is the idea of the energy cascade:

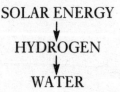

This is so simple that anyone who understands it once can never forget it.

- We use sunlight and convert it to electricity.
- Then we electrolyse water, which gives us hydrogen – the new clean fuel – and oxygen (which can be put back into the air or used).
- Hydrogen is then transported to population and industrial centres, using pipelines and tankers.
- At the population and industrial centres hydrogen is burned as a fuel, used in internal combustion engines in place of petrol and diesel, or used in fuel cells, thus giving mechanical power, heat or electricity.
- And the exhaust product from all these processes is water, the raw material we started with.

The solar-hydrogen system thus provides us with energy, but no resources other than sunlight are used, and there is no pollution.

WHAT HAS PREVENTED THE SPREAD OF THIS KNOWLEDGE?

There are several reasons. The most important is that the economics of the idea have only recently begun to look feasible.

The idea of hydrogen as an energy carrier dates from the 1970s, but until recently there was little optimism about the cost of the conversion of sunlight to the electricity necessary to electrolyse water. It has only been since 1989 that we have known that the material used for the photovoltaic cells – amorphous silicon – could be made cheaply enough for the final product, clean hydrogen, to be available at a cost less than that of petrol (and pollution).

Another reason why the processes described here have been limited to trials in research establishments is that, among the public, hydrogen is thought of as dangerous – not something you'd want to have fuelling your car or home. There is the image of the *Hindenburg* and the hydrogen bomb (despite the fact that the latter has absolutely nothing to do with what we are proposing).

But we think there might be another reason – maybe not as strong as the others, but still a part of the game: inertia. Many people don't really like changing to new things. The fuel

'I've heard about it — but why bother changing?'

stations are already there, and they sell fuel; it's so simple to go along, pay the money and drive off, just as we've done for decades. People think that all the damage done by traditional fuels is someone else's business – and responsibility to clear up – and nothing to do with them at all.

Last of all, and least of all right now, is the natural reaction of the producers of the present fuels. They didn't intentionally inflict on us the environmental ills that have been caused by their products. Indeed, the manufacturers have hardly

Just don't look up gentlemen ... keep your eyes on your profits.

accepted the greenhouse effect as yet, and when anyone points out that they are also causing acid rain – the decay of buildings, deforestation – cancer and so on, they look in the other direction. For them the most important thing is making a profit by selling a fuel already formed by nature.

We think that when people realise just how serious our environmental problems are they will insist on conversion to the new energy sources and clean fuels. This move may well not be greeted with great enthusiasm by the directors of Big Oil, Inc. – but then this is a world that is supposed to operate by the will and for the benefit of the people, not the Big Oil, Inc.

DEMOCRACY IN ACTION

A good example of the power of public pressure can be found in the United States where no nuclear reactors have been built since 1978.

In the 1950s, 1960s and 1970s, more than 90 nuclear reactors were built. However, by the 1970s, it became clear that nuclear reactors produced dangerous radioactive pollution, and although the degree of damage and the threat to health caused by pollution has always been controversial – some maintained that it was negligible – public concern began to grow.

The public probably also feared some kind of explosion and fire, the kind of disaster that actually occurred at Chernobyl in the Soviet Union in 1988, which spread radioactive debris and pollution over most of northern Europe.

So people started to take action. Lawsuits are very common in the United States, for instance you can take out a suit if a dog bites your child, or even if you suffer stress because of a noisy neighbour. Groups of people found they could join together and pool their economic resources as well as their common voice to get an injunction to stop the building of nuclear plants they didn't want in their community,

When courts issued injunctions, the companies building the reactors had to stop construction at the moment the injunction was delivered to them. Then the courts investigated the complaints – a process which often took several months.

Experts testified against one another and eventually the courts arrived at their decision. In many cases the decision went against the reactor which then remained unbuilt – a very expensive result for the companies building the plants. The people had won.

So the public in the USA proved they could do something to change decisions made over their heads, and the attitude that sometimes exists that Big Brother is there silently controlling everything is not always true. The case of the nuclear reactors has proved it, and this example of people power is particularly relevant to the subject of this book.

SPREADING THE NEWS

The authors of this book can't do much alone to spread the knowledge. Books are read by a few thousand people, of whom a mere few hundred retain the knowledge for more than a few months.

To make any headway in the democratic system there has to be political pressure. Our elected representatives and their aides have to know the views of the electorate. The most important thing a reader of this book can therefore do is to spread the news - that there is a cure for atmospheric pollution, acid rain and global warming, a new and revolutionary technology in which the sun's energy is collected and used to produce hydrogen, a totally clean and storable fuel.

And how is this information to be spread? Well, it will depend on your circumstances, on your enthusiasm, on your degree of contact with other people. For example, where do you work? In a factory, an office, a communications centre, a school, a night club? Wherever you work, there is the possibility of talking to other people. They won't want their atmosphere polluted, poisoned and destroyed, but they are unlikely to do anything about it unless there are viable alternatives to the present energy systems that deliver today's standard of living. You can explain what these alternatives are. More than that, you can explain that the initiative for change won't come from government; it has to come from you, the members of the public.

The next step is then to write/phone/pay a personal visit to your government representatives, who have to be re-elected every few years. They must be made aware that the people are willing to do something at the polls.

Furthermore, if, within a couple of years of the publication of this book, the knowledge of the energy cascade: sunlight - hydrogen - water - is in 20 per cent of all homes, that would make an enormous difference. If the public knows that a viable alternative is there, a market for the solar-hydrogen energy system is thus created. And once industries realise a market is there, they will invest their money and effort in creating what the people want.

THE POWER OF WOMEN

We've heard a great deal about equal rights in recent years, but the political clout of women is not exercised as much as it could, and should, be. This is particularly the case when it pertains to matters of health and the future.

We need to prepare for the future, a good future for ourselves, and particularly for our children and their children. We should be far less interested in the short term, for we can do little to affect the environment until the new technology we have described has been built. In this respect, women's actions are extremely relevant and helpful, and potentially very powerful when applied to the aim of permanently cleaning the environment. Women think about the future, and the future of their children.

But there is another reason why women should be more interested in promoting the idea that energy from the sun could be distributed as clean hydrogen fuel at costs less than our present fuels. Low-intensity nuclear radiation from some poorly functioning nuclear plants, or from nuclear accidents, affects a developing foetus more than it does the woman's body; the unborn child is extremely susceptible to such radiation, and in tragic cases it can result in physical abnormalities at birth. Furthermore, emissions from vehicle exhausts, especially carbon monoxide, affect the development of the brain in growing foetuses. These are just the sort of horrors that all expectant mothers fear. Up to now they have had no means of avoiding such exposure, except to say 'No' to the building of nuclear reactors, and to try and limit their

exposure to pollutants. But now the solar-hydrogen energy system gives them a positive clean alternative.

WRITING TO YOUR ELECTED OFFICIALS

Writing to your elected representatives is part of democracy; indeed, in many ways, apart from the ballot box, it's the main way democracy actually works. Elected representatives have aides and assistants, and one of their purposes is to help the elected officials know what the public is saying. Eventually, politicians have to respond to their electorate's wishes.

So, writing to the people you elect to office is good, helpful, and you can't do it too often. But keep the letter short and neat; one page is enough. Express a single idea per letter, and leave it at that. For example, you could suggest a carbon tax, or demand that polluters should pay.

ENDING THE UNFAIRNESS

We pride ourselves on being concerned with civil rights. But the truth is, air pollution violates everyone's civil rights. It robs us of liberty, for there is nowhere on this planet where we can go to be free of air pollution. Air pollution shortens life and it is unfair to all of us, but the ones who suffer most are the ones least able to defend themselves – children, the elderly and the poor.

Young children's immune systems are still developing, and are not able to ward off respiratory ailments caused by dirty air. The elderly suffer the same symptoms for the opposite reason; their immune systems are aging and unable to cope as well as they once did to ward off diseases.

Perhaps the most disturbing of all is the poor air that people in the inner cities and large suburban areas are forced to endure – buses and trucks belching diesel fumes; exhaust gases being emitted while taxis, delivery vans and other vehicles wait at traffic lights; the waste fumes of small industries, like dry cleaners and bakeries; and large industries, like utility plants and steel or paper mills, producing soot and ash. For this is where the poor and homeless live; they are the ones least able to afford to be sick with often chronic respiratory ailments caused by air pollution, and the ones least likely to afford the cost of housing outside the city.

It is time to end this unfairness. We advocate the implementing of an 'atmosphere users' fee' that would take into account the social, or real, cost of a product. These fees would then be collected and spent to develop clean, renewable energy for use in transport, manufacturing and in our own homes, thus giving everyone the quality of life to which they are entitled – air that is clean to breathe and an environment that is clean and safe to live in.

18
THE BOTTOM LINE

The preceding chapters have shown you that fossil fuels are the main culprits causing so much grief to the environment today; they cause air, water and soil pollution, acid rains and the greenhouse effect; they may be causing irreversible damage to the earth's environment and climate. We have also explained that the solar-hydrogen energy system is clean and environmentally compatible, and that both solar energy and hydrogen could be abundant and renewable. Once we make the conversion to the solar-hydrogen energy system, we will never have to convert to any other energy system.

The big question is: If the solar-hydrogen energy system is so wonderful, and fossil fuels cause so much grief, why aren't we using solar-produced hydrogen as our current fuel? The reason is that under the present economic system, fossil fuels don't have to pay for their environmental damage. The rules of the marketplace are biased toward fossil fuels and against alternative energy sources and fuels, even hydrogen.

What is the cost of fossil fuel-related illnesses, such as black lung, emphysema, lung cancer, skin cancer and others? What is the cost of the loss of timber, the loss of fish-bearing lakes and estuaries, the cost of a polluted ocean? The earth has a delicate ecosystem and can only clean up so much pollution itself. When the ecosystem is changed (by the greenhouse effect) and the natural chemical balances are affected (by acid rains), what means does the earth have to cleanse itself?

Already the oceans of the world are starting to show the results of our abuse; fish with blisters and cancer, dying of strange and unusual diseases; whales beaching themselves in increasing numbers; diminishing fishing catches; and so on. When the ocean becomes so polluted that we can't swim in it for fear of being poisoned, can we really expect the marine life that lives there not to be affected also? As we destroy the aquatic ecosystem, we are also dooming ourselves, because water, and the life in the water, is such an integral part of our lives.

So, what are the costs of these losses to human society? When we consider the frightening costs of the environmental damage caused by fossil fuels, and add them to the cost of the fuels themselves, we begin to understand the tremendous burden our planet has been carrying and how much it will cost to clean it up. We need to lighten the load.

Some predict that by the end of this century, or the beginning of the next, liquid fossil fuel (petroleum and natural gas) production will start to decline. We have to decide on an alternative energy system to take up this slack. As fossil fuels are also used as the raw materials for a wide range of products – lubricants, synthetic fabrics, aspirin and other medicines, inks, paints, to name but a few among thousands – we have to use what is left of these resources wisely so that future generations won't be deprived of them.

It will take until early into the 21st century for large-scale hydrogen production, distribution and storage systems to be constructed, and the equipment required for handling hydrogen to be designed and manufactured. And it will take several decades to convert the fuel stations, pipelines and tanker trucks needed for a full switch-over to the solar-hydrogen energy system. This is no bad thing, because the best way to proceed when changing our energy system is one step at a time; for example, it would cause financial chaos if we were suddenly to change from one fuel system to another. The sooner we make this conversion, however, the sooner we can stop and hopefully reverse the damage caused by fossil fuels.

ARE THERE ALTERNATIVES?

Now that you have read this far in the book, it must be obvious that we should convert to the solar-hydrogen energy system. But surprisingly there are many people who would like the present system to continue forever, even though sooner or later fossil fuels will run out. As oil and natural gas deposits are depleted, the fossil fuel advocates tell us that at first we can use coal deposits to manufacture synthetic petrol or gasoline (syngas) and synthetic natural gas (SNG), and then, when the coal too is depleted or too expensive, we can use shale oil and tar sands. Eventually, when there is not enough oil shale and tar sands to supply all our fuel needs, we could extract the carbon from limestone and use hydrogen from water, combining them to produce syngas and SNG. And then, when we run out of limestone (which would take quite a few mountains), they say we could take the carbon out of the carbon dioxide in the air and again use the hydrogen from water, and produce synthetic hydrocarbons. By these methods, say the fossil fuel advocates, we will have enough fossil fuel, or something very much like it, to last forever.

Those strongly in favour of fossil fuels don't mention that it would be extremely expensive to produce any syngas or SNG using such methods, nor do they mention the costs of continually repairing the environmental damage caused by these fuels. All these costs come to us in many different ways, some of them less obvious than others; trips to the doctor for a nagging cough that won't go away; or a new paint job for the car because it has so many rust spots from the corrosive action of polluted air and acid rain.

So, we could say that there are two main candidates for the post-fossil fuel era – synthetic fossil fuels (syngas, SNG) and the solar-hydrogen energy system. Therefore, we are going to compare these two possible systems by production costs, environmental damage and by how efficiently they can be used.

Production costs

Of course, the production costs of the synthetic fuels – syngas and SNG, and solar-produced hydrogen (hydrogen is also considered a synthetic fuel, but a non-polluting one) – must

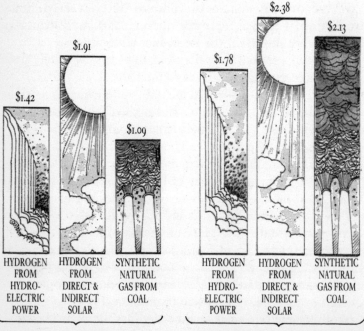

Synthetic fuel production
Costs for a gallon equivalent of petrol (1990 U.S.$)

be compared in terms of supplying worldwide energy demands.

As we read earlier, hydrogen can be produced using solar energy in various different ways – direct thermal, thermo-chemical, electrolytic and photolytic. But by using the direct and indirect forms of solar energy, the cheapest hydrogen we can produce now is from hydro-electric power (an indirect form of solar energy). To compare the costs of synthetic fossil fuels, we will only use those synthetics made from coal, since the ones made from shale oil and tar sands are much more expensive, and the synthetics made from limestone and hydrogen, and those using carbon dioxide from the air and hydrogen from the water would be even more so.

Among the gas synthetics, SNG is the cheapest, while among the liquid synthetics, liquid hydrogen from hydro-electric power is the cheapest (you may remember that liquid hydrogen is the preferred fuel for rocket and high-speed jet engines). By using this kind of chart (on page 134), it's easy to see which fuels are cheaper; but these amounts represent the cost of production only. Remember, in fossil fuels, even synthetic ones, there are hidden environmental costs. If we added them all in, how high would their columns extend?

Environmental damage
In calculating the costs of the fuels to society, their environ-mental effects and damages must certainly be considered; investigations are being conducted around the world to estimate these damages.

Fossil fuels are the main cause of damage to the biosphere (that part of this planet on which we live and breathe). The estimates for the year 1990 alone are that fossil fuels will spew out some 30 billion tons of carbon dioxide, other polluting gases, soot and ash. As we read in earlier chapters, these all produce pollution and acid rain and promote the greenhouse effect. Add to that the spills and leaks that occur during their storage and distribution, and the damage to people and natural environment becomes overwhelming.

Another form of damage caused by fossil fuels is the ugly, denuded landscape left by the strip-mining of coal. It has been reported that the United States government alone spent US$19 billion in the period 1980 to 1989 for land reclamation to correct some of the damage caused by strip-mining. At the same time industry spent an equivalent amount.

When agricultural land becomes unfit for cultivation because of pollution, this results in additional losses to society. In other words, the loss cannot be calculated only in terms of the replacement costs of agricultural soil; the time for which the use of the land is not available must also be taken into account.

Reports have shown that ozone at ground level, formed by the interaction of exhaust fumes with sunlight, reduces the sugar content in grapes and therefore the subsequent wine yields. It is also estimated that ozone is responsible for the loss of 5 per cent of the annual farm produce in the United States, which is enough to meet the food needs of Brazil and Mexico put together.

The destructive effects of pollution on forests are already noticeable. As acid rain, snow, sleet or other precipitation soaks into the soil, it dissolves metals and other minerals that occur naturally and are not normally dissolved by ordinary rainwater. These dissolved substances, particularly those containing aluminium, are toxic to the roots of young trees. Atmospheric pollution is also harmful to animals as well as human beings.

Then there is the damage to the sea. Oil spills in seas and oceans by various kinds of ship, and particularly those spills caused by tanker accidents, pollute the oceans and the shores, causing billions of dollars in damage. The tanker *Amoco Cadiz*, back in 1978, ran aground off the coast of Brittany while carrying 60 million gallons of crude oil. The resulting oil spill caused damages in the region of US$1.5 billion. Another crash in November 1982, of the *Globe Asimi*, carrying 4.8 million gallons of fuel oil off the Baltic Sea coast of the Soviet Union, is reported to have caused damages of about US$900 million. The more recent grounding of the *Exxon Valdez* off the port of Valdez, Alaska, in 1989, has been estimated to have caused US$11 billion worth of damage, and who knows the cost of the vast quantities of oil spilt into the Persian Gulf during the Gulf war in 1991. Accidents occurring at offshore oil rigs produce substantial pollution, too, while ballast (or bilge) water, habitually discharged into the sea by tankers, adds continuously to the pollution.

But the damage caused by pollution and all the problems this pollution inflicts on the earth is occurring so slowly that many of us are missing the most devastating change of the century – the sea levels are rising. This fact has been established by data collected worldwide, and is caused by the melting of the polar ice caps and glaciers. And the increase in temperature that causes this melting is a result of the greenhouse effect, which in turn is directly attributable to the increasing amounts of carbon dioxide in the atmosphere. It has been estimated that the present rate of ocean rise is about 1 cm (half an inch) per year; however, it is likely to increase because any significant warming and subsequent rise in sea level would trigger the rapid break-up and melting of the vast Antarctic ice sheet. The damage caused by the rising oceans will be huge, since it will involve some of the largest, most densely populated and commercial cities around the world, as well as the more fertile lands, such as the great river deltas in Egypt, Bangladesh and Louisiana.

The greenhouse effect, caused by the increasing amounts of carbon dioxide and other gases in the atmosphere, may have other serious consequences, too. Rising temperatures would eventually result in the expansion of existing deserts. The present temperate regions will become subtropical, while the colder climates will become temperate. This in turn will push the agricultural lands north in the northern hemisphere and south in the southern hemisphere into areas which may be less fertile or unsuitable for agriculture. As a result, the amount of agricultural land worldwide will be smaller, while the world's population continues to grow.

Pollution can have many insidious yet dangerous effects on human beings. Although human diseases caused by pollution and acid rain have been studied in terms of medical care and treatment, those expenses don't cover mental damage, human discomfort and the general unhappiness involved. People are unhappy when they are in pain, and unhappy workers are less productive. Nobody has even given an estimate of these kinds of costs.

Another factor which has not yet been accounted for is the cost of protecting oil supplies in political crises. For example, the cost of the Gulf war of 1991 will not be met by the oil companies but by people's taxes.

The diagram opposite shows how much environmental damage costs (1990 US dollars). However, this is not an all-inclusive list by any means. The total estimated damage for only the items considered comes to US$1.30 per gallon of oil, or equivalent in other fossil fuels. Don't forget that this damage will eventually be paid for by society in general. If we do not take corrective action now, the coming generations will have to pay a much greater price to undo the damage. And some of the damage may never be undone.

This result is very interesting. It tells us that if we take the average cost of petrol US$1.00 per gallon, its cost to society would actually be US$2.30 per gallon. So even today, the cost of solar-hydrogen would be less (by 40 per cent) than the cost of petroleum.

How efficient they are

In comparing the synthetic fossil fuels with hydrogen, it is important to compare the efficiencies of these products for consumer use. For the consumer, fuels should produce various forms of energy, such as mechanical in a car, electrical in a power plant, and heat in a cooker. Studies show that in almost every application, hydrogen can be converted to the desired energy forms more efficiently than fossil fuels or synthetic fossil fuels. In other words, using hydrogen as a fuel would result in energy conservation due to its higher efficiency.

On the left-hand side of the chart on page 140, we start off with 1,000 units of synthetic fossil fuel energy to meet transport and commercial, industrial and residential demands. Compare this to the right-hand side which shows the units of hydrogen necessary to produce the same amounts of energy needed to satisfy the same demand for energy. Because we need much less hydrogen, it is much more efficient.

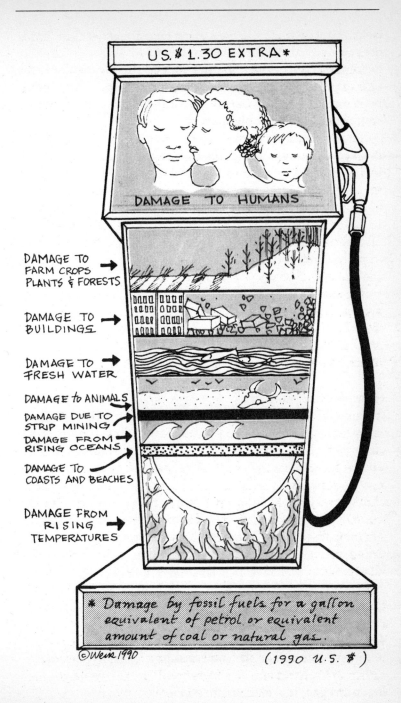

U.S. $ 1.30 EXTRA *

DAMAGE TO HUMANS

DAMAGE TO
FARM CROPS
PLANTS & FORESTS

DAMAGE TO
BUILDINGS

DAMAGE TO
FRESH WATER

DAMAGE to ANIMALS

DAMAGE DUE TO
STRIP MINING

DAMAGE FROM
RISING OCEANS

DAMAGE TO
COASTS AND BEACHES

DAMAGE FROM
RISING
TEMPERATURES

* Damage by fossil fuels for a gallon
equivalent of petrol or equivalent
amount of coal or natural gas.

© Weir 1990

(1990 U.S. $)

Cost to society

From the last three sections we get a good idea of the cost to society of each of the synthetic fuels. In the case of the synthetic fossil fuels we know we have to add the cost of environmental damage to the cost of production in order to understand the real expense of using fossil fuels, even synthetic ones.

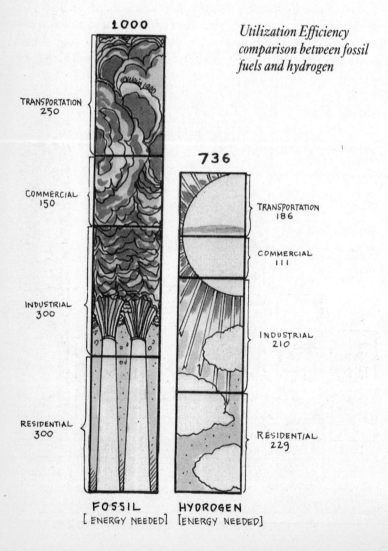

Utilization Efficiency comparison between fossil fuels and hydrogen

1000

TRANSPORTATION 250

COMMERCIAL 150

INDUSTRIAL 300

RESIDENTIAL 300

736

TRANSPORTATION 186

COMMERCIAL 111

INDUSTRIAL 210

RESIDENTIAL 229

FOSSIL
[ENERGY NEEDED]

HYDROGEN
[ENERGY NEEDED]

In the case of hydrogen produced from direct and indirect forms of solar energy, there is no environmental damage or cost caused by either the sunlight or by hydrogen, the energy go-between. Of course, the fact that hydrogen is more efficient must also be taken into account. It is about 26 per cent, more efficient than both liquid and gas; in other words, to get the same amount of work from something that requires a fuel, we need 26 per cent less hydrogen than a fossil fuel.

If we were to measure hydrogen gas in gallons or litres, as we do with our current fuels, then hydrogen produced from hydro-electric power would be the cheapest in real costs. as seen in the figure below SNG's cost to society is more than

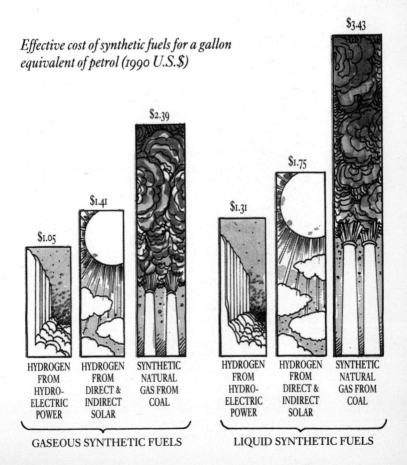

Effective cost of synthetic fuels for a gallon equivalent of petrol (1990 U.S.$)

$1.05 — HYDROGEN FROM HYDRO-ELECTRIC POWER

$1.41 — HYDROGEN FROM DIRECT & INDIRECT SOLAR

$2.39 — SYNTHETIC NATURAL GAS FROM COAL

GASEOUS SYNTHETIC FUELS

$1.31 — HYDROGEN FROM HYDRO-ELECTRIC POWER

$1.75 — HYDROGEN FROM DIRECT & INDIRECT SOLAR

$3.43 — SYNTHETIC NATURAL GAS FROM COAL

LIQUID SYNTHETIC FUELS

twice that of hydro-electric power-produced hydrogen; its cost is also greate (by about 70 per cent) than hydrogen produced by either direct solar energy or by indirect forms of solar energy other than hydro-electric power. When we look at liquid fueld, again liquid hydrogen from hydro-electric power is the cheapest, followed by liquid hydrogen from other solar sources. Syngas is more than two and a half times more costly to society as hydro-electric power hydrogen, and about twice as costly as hydrogen from other solar sources.

Another way of thinking about the social cost of a fuel is to consider possible future energy scenarios. We could continue with the present fossil fuel system by meeting our energy needs half with syngas and half with SNG. Syngas could be used in cars, planes, trains and so on, and SNG would be used where we use natural gas today, like heating and cooking.

Another scenario would be the solar-hydrogen energy system, in which we assume that 25 per cent of the energy needs would be met by liquid hydrogen – for example, the air and space transport – and the rest would be met by hydrogen gas. Let's say that one-third of the hydrogen would be

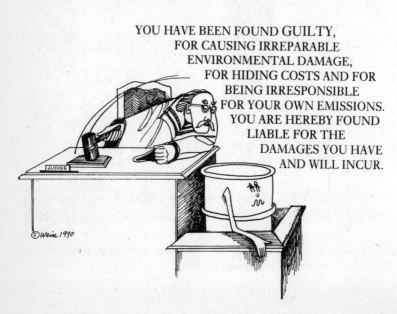

YOU HAVE BEEN FOUND GUILTY, FOR CAUSING IRREPARABLE ENVIRONMENTAL DAMAGE, FOR HIDING COSTS AND FOR BEING IRRESPONSIBLE FOR YOUR OWN EMISSIONS. YOU ARE HEREBY FOUND LIABLE FOR THE DAMAGES YOU HAVE AND WILL INCUR.

produced by hydro-electric power and the remainder by direct and other indirect solar sources; the chart opposite shows how the costs change when their real costs are taken into account.

We can see that the cost of the synthetic fossil scenario to society would be equivalent to US$2.91 per gallon. In the case of solar hydrogen, the cost would be US$1.37. In other words,

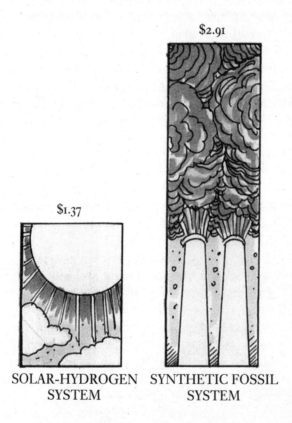

$2.91

$1.37

SOLAR-HYDROGEN SYSTEM

SYNTHETIC FOSSIL SYSTEM

Comparison of Solar-Hydrogen and synthetic fossil fuels scenarios: effective cost of energy for a gallon equivalent of petrol (1990 U.S.$)

the syngas and SNG system would cost more than double that of the solar-hydrogen energy system.

ENERGY-ENVIRONMENT RESOLUTION

When the advantages of the solar-hydrogen energy system are

brought to the attention of decision-makers, legislators and government officials, their answer is usually 'Hydrogen sounds good, but it is more expensive than petroleum. Let the open market decide what the energy system should be.' That answer is all right if the rules of competition are fair, but at present they are not. The rules favour the lower production costs of petroleum, irrespective of its environmental effects. We need new and fair laws that take into account not just the production costs but also the cost of the environmental damage. Fossil fuels must be liable for their own damages.

In order to clarify these ideas, some 65 energy and environmental scientists from around the world, headed by Professors Veziroglu and Bockris, have drafted an 'Energy-Environment Resolution'. It is reprinted here for you to consider.

ENERGY-ENVIRONMENT RESOLUTION

Whereas the countries of the world are striving to increase their living standards, and thus – in addition to adopting all appropriate conservation measures – must increase their energy consumptions in the long run,

Whereas the main energy sources at present are fossil fuels,

Whereas fossil fuels are finite in amount and will eventually be depleted, with the downturn in production expected to start early in the next century, if not before,

Whereas it is prudent to plan and begin conversion to the next energy system by making use of the remaining fossil fuel sources (and also other conventional energy sources, such as nuclear, etc.) to achieve a smooth changeover, which is expected to take half-a-century or so,

Whereas it is also prudent to preserve the diminishing supplies of the fossil fuels for non-fuel applications (such as lubricants, synthetic fibres, plastics and fertilizers), for which there may be no substitutes,

Whereas the combustion products of fossil fuels are causing growing damage to our Biosphere (the only known domain in the Universe to be supportive of life) and especially to its living components through pollution, acid rain, CO_2 and carcinogens,

Whereas the combustion products and their harmful effects do not stop at the national boundaries,

Whereas it is of the utmost importance to keep the Biosphere clean and fit for life, and hence the energy sources and energy carriers as clean as possible,

Whereas there exist 'clean' primary energy sources,

Whereas there is a need for two types of energy carriers, viz., electricity (meeting about one quarter of the demand at the consumer end) and fuel (meeting about three quarters of the demand),

Whereas there exists technology for the production and utilisation of the environmentally most compatible and most efficient fuel energy carrier, i.e., hydrogen,

Whereas there is a necessity for a *self-regulating system*, to ensure that man-induced factors (energy or otherwise) do not harm the Biosphere or threaten life,

It is hereby resolved:

That products (energy carriers and otherwise) be made responsible for the harm they cause to the Biosphere and to life, directly or through their waste or their manufacture,

That the price of each product include an 'environmental surcharge' to cover its environmental damage,

That the environmental surcharge be used by appropriate authorities to undo the damage to the Biosphere, life and structures; to cover related medical, restoration and relief expenses; and to compensate the victims,

That there must be international cooperation to ensure equitable and uniform application of the environmental surcharge.

Implementation of the above

Will result in the following lastings and universal benefits:

Ensure that eventually the present fossil fuel system would be replaced by a clean and environmentally compatible energy system, the Hydrogen Energy System,

Ensure that energy (fossil, nuclear, etc.) would be available for the changeover, thus resulting in a smooth change with no upheavals,

Ensure that some supplies of fossil fuels would be preserved for non-fuel applications, for which there might be no substitutes,

Promote worldwide economic development and hence international harmony as progressively less unproductive work would be required to undo the environmental damage, and through the utilisation of an efficient and renewable energy carrier.

Save the Biosphere and life from extinction by ensuring that products, their manufacturing methods and wastes would be environmentally compatible.

The principle expressed in the resolution has already been applied on a small scale; for example, if the pollution from a factory is damaging farms around it, then the farmers are compensated. What we are proposing is that this principle should be applied universally, and not only for immediate damage like a gas pipeline leak or an oil spill, but also for the long-term damage that may take place far away from the source and over a period of time, like the loss of sugar maple trees from acid rain. Under such a system, when calculating the price of a product, a surcharge would be added to cover the cost of the environmental damage caused by its production and manufacture. This surcharge could then be used, by the appropriate authorities, to undo the damage to the biosphere caused by the pollutants, to cover related medical, restoration and relief expenses, and to compensate the victims. If the 'polluter pays' principle is made law, there will be a much greater incentive to invest in clean, safe fuels, and the solar-hydrogen energy system is the obvious choice. But there will be other important benefits. Eventually, it will force all products to be clean and environmentally compatible; it will save the biosphere from slow death and preserve the life on this planet; and also provide a higher quality of life for its inhabitants - humans, animals and plants.

We therefore ask you, our readers, the public at large, to urge your legislators and your government officials to pass laws containing the principles expressed in the resolution. You must explain that:

- It will protect the environment.

- It will protect the public health.
- It will reduce wasteful spending.
- It will provide built-in incentives for goods and production methods to be environmentally compatible.
- And it will help reduce consumer costs, which will often mean reduction in taxes, since the unproductive spending on correcting environmental damage and on pollution-related health care will progressively be phased out.

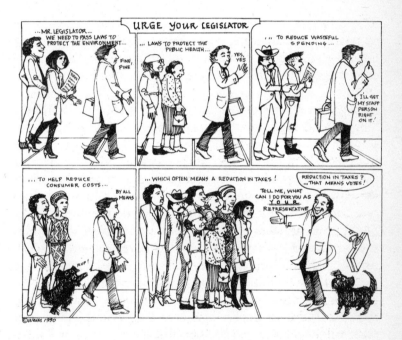

The solar-hydrogen energy system is an idea whose time has arrived. There can be no going back. We have no doubt it will prevail, and the good earth will have the energy system it deserves – a system that is supportive of its ecosystem. So you see, it is all up to you.

All Optima books are available at your bookshop or newsagent, or can be ordered from the following address:

Optima, Cash Sales Department,
PO Box 11, Falmouth, Cornwall TR10 9EN

Please send cheque or postal order (no currency), and allow 60p for postage and packing for the first book, plus 25p for the second book and 15p for each additional book ordered up to a maximum charge of £1.90 in the UK.

Customers in Eire and BFPO please allow 60p for the first book, 25p for the second book plus 15p per copy for the next 7 books, thereafter 9p per book.

Overseas customers please allow £1.25 for postage and packing for the first book and 28p per copy for each additional book.